Molecular Genetics

Peter Smith-Keary

MACMILLAN

First published 1991

Published by
MACMILLAN EDUCATION LTD
Houndmills, Basingstoke, Hampshire RG21 2XS
and London
Companies and representatives
throughout the world

Printed in Hong Kong

British Library Cataloguing in Publication Data
Smith-Keary, P. F. (Peter F.)
Molecular genetics.
1. Molecular genetics
I. Title
574.87328
ISBN 0–333–52978–2

Contents

Introduction

This book is intended as a revision manual for students taking first-year university courses in Molecular Genetics or Genetics and is particularly directed at those intending to pursue further courses in Molecular Genetics and/or Molecular Biology. It is not intended that it should replace a standard introductory text but rather that one should complement the other, and for this reason many features of a textbook, such as historical perspective, detailed descriptions of experimental techniques, multiple examples and lists of further reading, are not included. However, all the basic information required for most introductory courses in Molecular Genetics is included.

The first 11 chapters contain material that is common to nearly all introductory courses in Genetics and Molecular Genetics. Chapter 12 introduces some highly selected aspects of the molecular genetics of eucaryotes, and Chapter 13 outlines some fundamental aspects of genetic engineering; the material in these chapters will be relevant to some courses but not to others.

The Lay-out of the Book

Each chapter is in three parts:

(1) A concise description of basic concepts.
(2) A wide range of specimen answers to examination-style questions. Considerable emphasis is placed on these answers and, when using this book for revision, it is important to appreciate that most of the detailed information together with any more advanced material is *only* given in these answers; thus, these model answers are a fundamental part of the text.
(3) A short selection of further questions, brief answers to which are given at the end of the book.

The Questions and Answers

The questions require answers of varied length, ranging from a few lines to a short essay (30–40 minutes in an examination), but, to increase their usefulness, some of the longer questions are sectionalised and so can be regarded as several shorter questions and answers. Most of the text itself is also written in the style of answers to questions suggested by the chapter sub-titles.

Since the answers are a fundamental part of the text, it has frequently been necessary to provide some information and explanation additional to that contained in the examination-style answers; this further material, which would not be relevant to an examination answer, either is enclosed in parentheses or appears as notes following the answer.

For convenience in cross-referencing and to increase their use for revision, the figures in the specimen answers have both serial numbers and descriptive legends; these would not normally be required in an examination answer.

The student is strongly recommended to read each question and to plan carefully his or her answer before proceeding to read the model answers.

Answering Examination Questions

Three particular points should always be adhered to:

(1) *Read the question carefully* It is surprising how frequently students misread a question and so fail to answer the question as set.

(2) *Keep to the point* Do not waste time by including irrelevant material. If, for example, you are asked to describe how DNA replicates, it is quite pointless to include a description of DNA structure; you will ONLY be marked on what is relevant to the question as set.

(3) *Apportion your time* This is necessary so that you have time to attempt all the questions that you are required to answer. If you answer less than the required number of questions, you will get one or more zero marks.

Good luck to all who use this book!

Leominster, 1990   P.S.-K.

Glossary

Alleles Alternative forms of the same gene (for example, *vg* and *vg*$^+$). (47)

Anticodon The sequence of three nucleotides in a molecule of tRNA that recognises a complementary sequence of nucleotides (the codon) on a molecule of messenger RNA. (132)

Antisense strand The strand of the DNA duplex that acts as the template for messenger RNA synthesis. (130)

Autoradiograph The photographic image produced by labelling a molecule, such as a phage or bacterial chromosome, with a radioactive label and overlaying it with a photographic emulsion. The emissions from the radioactive label expose the emulsion and, after development, reveal the size and shape of the underlying molecule. (36)

Autosome Any chromosome that is not directly involved in the determination of sex. (7)

Auxotroph A mutant micro-organism that can only grow when a particular amino acid, nucleotide or vitamin is provided in the medium. (98, 113)

Back-cross A cross between an organism and one of its parents. (49)

Back-mutation A mutation which results in a mutant gene regaining its normal activity by restoring the exact nucleotide sequence present in the wild-type gene. (94, 192, 249)

Bacteriophage A virus that infects bacteria. (9, 88)

Base substitution The replacement of one base (or base pair) in a nucleic acid by a different base (or base pair). (162, 183)

Bivalent The pair of synapsed homologous chromosomes present during the prophase of meiosis. (5)

Carcinogen A physical or chemical agent that causes cancer. (190)

Centromere A specialised region of a chromosome to which the spindle fibres attach at cell division. (3, 35)

Chain-termination (nonsense) triplet (CTT) A codon signalling the termination of polypeptide synthesis. These codons are UAG (amber), UAA (ochre) and UGA (opal). (135, 185)

Chiasma The crossed-over strands of two non-sister chromatids seen at the first meiotic division in eucaryotes; the position at which two homologous chromosomes appear to exchange genetic material during meiosis. (5, 11)

Chromatid In eucaryotes one of the two identical strands of a newly replicated chromosome. (3)

Chromatin The nucleoprotein complexes that form the chromosomes. (33)

Chromomere A laterally differentiated region of a chromosome. In meiotic chromosomes the chromomeres appear like beads on a string but in the giant salivary gland chromosomes each chromomere is seen as a transverse band. (39)

Chromosome The molecule of nucleic acid (in procaryotes and viruses) or the complex of DNA, RNA and protein (in eucaryotes) carrying the genetic information. (2, 31, 33)

***Cis*-arrangement** Describes the situation where two mutant sites or genes are on the same chromosome or molecule of DNA; in eucaryotes a double heterozygote in the coupling phase (*AB/ab*). (115)

***Cis–trans* test** A genetic test to determine whether two mutations are in the same or in different genes. (115)

Cloning The production of very many genetically identical molecules of DNA, cells or organisms. (223)

Codon The three consecutive nucleotides in RNA or DNA encoding a particular amino acid or signalling the termination of polypeptide synthesis. (151)

Cohesive (sticky) ends Single-stranded and base-complementary sequences of nucleotides at opposite ends of a molecule of DNA. (42)

Complementation The ability of two mutant genes to make good each other's defects when present in the same cell but on different molecules of DNA. (114)

Concatamer An end-to-end (tandem) array of identical DNA molecules. (29)

Conditional lethal mutant A mutant able to grow under one set (permissive) of environmental conditions but lethal under different (restrictive) conditions. (183)

Consensus sequence An idealised nucleotide sequence where the base at each particular position is the one most frequently observed when many different actual sequences are compared. (138)

Constitutive gene expression A gene or operon which is expressed at all times and under any environmental conditions. (169, 171)

Coupling Two pairs of linked genes are in the coupling phase when both dominant alleles were contributed by one parent and both recessive alleles by the other parent (for example, *AB/ab*). (75)

Crossing-over The process of genetic recombination that gives rise to new combinations of linked genes. (10, 65)

Deletion The loss of one or more base pairs from a molecule of DNA. (152, 185)

Denaturation The treatment of a molecule of double-stranded DNA so that it separates into its two component strands. (27, 141)

Dihybrid cross A cross between two dihybrids (for example *A/a B/b* × *A/a B/b*). (51)

Diploid A eucaryotic cell or organism in which the chromosomes exist in pairs. (3)

Direct repeats Two identical (or nearly identical) nucleotide sequences sometimes separated by a sequence of non-repeated DNA, for example
5′ ATTCGA........ATTCGA 3′
3′ TAAGCT ....TAAGCT 5′ (199)

DNA ligase The enzyme that joins together the 5′ and 3′ ends of polynucleotide chains by forming a phosphodiester bond between them. (22)

DNA polymerases The enzymes that polymerise deoxyribonucleotides on to existing polynucleotide chains using the complementary DNA strands as templates. (21)

DNA primase The enzyme that normally synthesises the RNA primers required for initiating DNA synthesis. (22)

Dominant gene The gene (or, more correctly, the character) that is expressed in a heterozygous or partially heterozygous cell (for example, vg^+ in vg^+/vg). (48)

Endonuclease An enzyme that makes breaks in a molecule of DNA by hydrolysing internal phosphodiester bonds. (237)

Eucaryote An organism with nuclear membranes and certain organelles, such as mitochondria, and mitotic spindles. (1)

Euchromatin The parts of a eucaryotic chromosome that show the normal cycle of condensation at cell division. (33)

Exonuclease An enzyme that digests a molecule of nucleic acid by removing successive nucleotides from the 5′ or 3′ end. (21, 237)

F$^+$ cell A bacterial cell harbouring an F plasmid. (91)

F plasmid or **F factor** The fertility factor of *Escherichia coli*; it enables cells to conjugate. (91)

F-prime plasmid (F′) An F plasmid carrying a segment of the bacterial chromosome. (95)

Frameshift A mutation which adds or deletes one or two base pairs from a coding sequence in a molecule of DNA, so that the genetic code is read out-of-phase. (152, 185)

Gene The genetic unit of function; it may (i) encode a polypeptide, (ii) encode a molecule of ribosomal or transfer RNA or (iii) be a regulatory sequence involved in the control of gene expression. (47)

Genome The complete gene content of a cell or organism.

Genotype A specific description of the genetic constitution of an organism. (48)

Haploid A cell or organism with only one set of chromosomes. (3)

Hemizygous Describes the situation in a diploid organism whereby a certain gene or chromosome is only present in a single copy, for example, the X-linked genes and the X chromosome in an XY male. (54)

Heterochromatin Chromatin that is relatively over- or under-condensed at cell division. (34)

Heteroduplex A molecule of double-stranded nucleic acid where the two strands are of different origin and so may not have exactly complementary base sequences. (208)

Heterozygote A cell or organism where different alleles are carried by the two members of a pair of homologous chromosomes. (3, 47)

Histones Basic proteins found complexed with DNA in eucaryotic chromosomes. (33)

Homozygote A cell or organism where the same allele is carried by each member of a pair of homologous chromosomes. (3, 47)

Inducible system A regulatory system where the genes are only turned on and enzymes synthesised when the appropriate substrate is present. (165)

Insertion sequence (IS) A transposable nucleotide sequence that only encodes functions related to its own transposition. (190, 198)

Inverted repeats (IR) Where the sequence of nucleotides along one strand of DNA is repeated in the opposite physical direction along the other strand; inverted repeats are often separated by a tract of non-repeated DNA. For example

$$5′ \quad \text{GATG ..... CATC} \quad 3′$$
$$3′ \quad \text{CTAC ..... GTAG} \quad 5′ \quad (23)$$

Karyotype A description of the chromosome content of a cell or organism. (7)

Kilobase (kb) A unit of length of 1000 nucleotides or 1000 nucleotide pairs.

Lagging strand The strand of newly replicated DNA that is synthesised discontinuously and away from the replication fork. (20)

Leading strand The strand of newly replicated DNA that is synthesised continuously and towards the replication fork. (20)

Ligation The formation of a phosphodiester bond between two adjacent bases separated by a single-strand break. (42)

Linkage The tendency of genes close together on the same molecule of DNA (chromosome) to be inherited together. (65)

Linkage map A map, assembled from recombination data, showing the order of mutant sites and genes along a chromosome. (66)

Locus The site on a chromosome where a particular gene is located. (3)

Lysogen A bacterial cell carrying a phage genome as a repressed prophage. (88)

Lytic response When an infecting phage genome replicates, matures and eventually lyses the bacterial cell, releasing free phage. (88)

Meiosis The special type of cell division, which halves the number of chromosomes in a diploid organism so as to form haploid gametes or spores. (4)

Messenger RNA (mRNA) The transcript of a segment of chromosomal DNA which is a template for polypeptide synthesis. (130)

Micrometre (μm) 1×10^{-6} metres, 1000 nm.

Missense mutation A mutation that changes a codon for an amino acid into a codon for a different amino acid. (184)

Mitosis The normal process of cell division. (3)

Mutagen An agent that increases the frequency of mutation. (186)

Mutant A cell or organism with a defective gene and displaying a specific altered phenotype. (47, 182)

Mutation The process that produces a sudden heritable change in the nucleotide sequence of an organism; any such change in the nucleotide sequence. (182)

Nanometer (nm) 1×10^{-9} metres.

Negative control A regulatory system where gene activity only occurs in the *absence* of a specific regulatory protein. (169)

Nonsense mutation A mutation which alters a codon for an amino acid to a chain-termination triplet. (155, 185)

Nucleolus A granular structure found in eucaryotic cells attached to a specific chromosomal site, the **nucleolar organiser**. It is the site of ribosomal RNA synthesis. (2)

Nucleoside A purine or pyrimidine base linked to a molecule of ribose or deoxyribose. (15)

Nucleotide A nucleoside with an attached phosphate group. (16)

Okazaki fragments The short discontinuously synthesised fragments that are ligated together to form the lagging strand during DNA replication. (20)

Operator The DNA sequence to which a repressor protein reversibly binds so as to regulate the activity of one or more structural genes. (167)

Operon A group of co-ordinately controlled structural genes, together with the operator and promoter sequences that control them. (166)

Palindrome A pair of adjacent inverted repeat sequences. (224)

Phage (bacteriophage) A virus that infects a bacterial cell. (9, 88)

Phenotype The appearance or other observable characteristics of an organism. (48)

Phosphodiester bond The covalent bond joining the 3′-hydroxyl of the sugar moiety of one (deoxy) ribonucleotide to the 5′ hydroxyl of the adjacent sugar. (17)

Plasmid An extra-chromosomal, covalently closed circular molecule of DNA carrying non-essential genetic information and replicating independently of the chromosomal DNA. (91)

Point mutation A mutation involving the addition, deletion or substitution of a single base pair (or, occasionally, several adjacent base pairs). (183)

Polypeptide A compound made up of two or more amino acids joined together by peptide bonds. (109)

Positive control A regulatory system where gene activity only occurs in the *presence* of a gene-encoded regulatory protein. (170)

Primer The short strand of RNA which provides the free 3′—OH end on to which DNA polymerase can then add deoxyribonucleotides. (20)

Procaryotes Organisms lacking nuclear membranes and certain organelles, such as mitochondria, and mitotic spindles. (8)

Promoter The DNA sequence to which RNA polymerase binds in order to initiate transcription. (130, 138)

Prophage A phage genome whose lytic functions are repressed and which replicates in harmony with the bacterial chromosome; it is, as with lambda, frequently integrated into the chromosome of the host bacterium. (88)

Prototroph A micro-organism able to grow on minimal medium containing only certain inorganic salts and a carbon source. (113)

Reading frame One of the three ways by which a coding sequence of nucleotides can be read in consecutive groups of three. (152)

Recessive The gene (or, more correctly, the character) that is not expressed in a heterozygote (for example, *vg* in *vg⁺/vg*). (48)

Recombinant DNA A DNA molecule made by joining together two different DNA molecules. This term is usually applied to molecules made by *in vitro* ligation. (225)

Recombination The process which produces new combinations of the genetic material following the occurrence of crossing-over between two homologous molecules of DNA. (65)

Redundant Applies to a gene or nucleotide sequence that is present in more than one (usually many) copies. (48, 204)

Replication fork The Y-shaped region of a DNA molecule where the two strands have separated and replication is taking place. (19, 21)

Replicon A molecule of DNA able to initiate its own replication.

Repressor The protein product of a regulator gene; when it binds to its own operator it prevents transcription of the adjacent structural genes. (168)

Repulsion Two pairs of linked genes are in the repulsion phase when the paternal and maternal chromosomes each carry one dominant and one recessive gene (for example, *Ab/aB*). (75)

Restriction endonuclease A large group of endonucleases each of which recognises and attacks a specific DNA sequence; they are extensively used in recombinant DNA technology. (224)

Reversion Any mutation that restores (or partially restores) the wild phenotype of an organism. (158, 191)

RNA polymerase The enzyme that normally synthesises RNA against a DNA template. (23)

RNase An enzyme that hydrolyses RNA molecules. (237)

Rolling-circle (sigma) replication The type of replication where a replication fork moves round and round a circular molecule of DNA producing a single-stranded concatamer which may then become double stranded by the synthesis of a complementary strand. (29)

Same-sense mutation A mutation which changes the nucleotide sequence of a codon so that it still encodes the same amino acid. (184)

Screening A technique to increase the proportion of a rare phenotype in a population so making its recovery easier. (191)

Selection A technique, using a specific set of environmental conditions, which allows the survival of mutant or recombinant cells having a particular phenotype and excludes any other more common phenotypes. (90, 191)

Semi-conservative replication The method whereby DNA normally replicates so as to produce two daughter molecules each consisting of one parental and one newly synthesised strand. (19)

Sense strand The strand of the DNA duplex that is *not* transcribed and which has the same nucleotide sequence as the messenger RNA. (131)

Silent mutation A mutation which changes the nucleotide sequence but which has no detectable effect on the phenotype. (182)

Sister chromatids In eucaryotes, two identical chromatids held together by a common centromere. (3)

Site The position of a mutation within a gene; a specific base pair. (95, 196)

Suppressor mutation A mutation that restores, or partially restores, the loss of function caused by a different mutation. Many suppressor mutations are in genes encoding a species of tRNA; the altered tRNA can recognise the original mutant codon and, during translation, insert an acceptable substitute amino acid into the polypeptide. (193)

Tandem duplication A directly repeated DNA sequence. (211)

Temperate phage A phage that can either establish itself as prophage or can enter the lytic response when it infects a sensitive cell. (88, 99)

Terminator The DNA sequence that signals the end of transcription. (131, 139)

Tetrad The four products of a single meiosis. (6, 73)

Theta-type (Cairns-type) replication Replication of a circular molecule of double-stranded DNA by initiation at a unique origin and proceeding in one or both directions around the circular molecule. (32)

***Trans*-arrangement** Describes the situation where two mutant sites or genes are on different chromosomes or molecules of DNA; in eucaryotes a double heterozygote in the repulsion phase (*Ab/aB*). (114)

Transcription The process by which RNA polymerase (or DNA primase) produces a molecule of RNA using one strand of a DNA duplex as a template. (130)

Transduction The transfer of genes from one bacterium to another by a phage vector. (93)

Transfer RNA (tRNA) Adaptor molecules which, on the one hand, bind to a particular amino acid and, on the other hand, recognise the corresponding amino acid codon on the messenger RNA, thus correctly aligning the amino acids for assembly into a polypeptide. (132)

Transformation The transfer of genes from one bacterium to another by 'infecting' a recipient strain with purified DNA extracted from a genetically different donor. (26, 95)

Translation   The assembly of amino acids into polypeptides using the genetic information encoded in the molecules of messenger RNA. (131)

Transposon   A transposable genetic element which, in addition to encoding the proteins required for its own transposition, confers one or more new recognisable phenotypes (usually resistance to one or more drugs) on the host cell. (190)

Transposition   The transfer of a discrete segment of DNA from one location in the genome to another. (190)

Vector   An independent replicon, usually a small plasmid or viral genome, used to introduce a 'foreign' gene into a host cell. (223)

Virulent phage   A phage that can only enter the lytic cycle when it infects a sensitive host cell. (88, 99)

Zygote   The diploid formed by the fusion of the male and female gametes. (4)

*The numbers refer to the pages where the definitions are more fully explained. The explanations of technical terms not listed in this glossary can be found in the text by reference to the index.

1 Cells and chromosomes

1.1 Eucaryotic Cells

In all organisms, from bacteria to man, the fundamental unit of structure is the **cell**. A typical animal cell (Figure 1.1) is bounded by a semipermeable **cell** or **plasma membrane** which acts as a barrier between the exterior and the interior of the cell and controls the movement of substances in to and out of the cell. This membrane encloses the **cytoplasm,** within which are suspended a variety of membrane-bound compartments called **organelles**; the cytoplasm also contains a matrix of filaments and tubules, the **cytoskeleton,** which provides structural support for the cell.

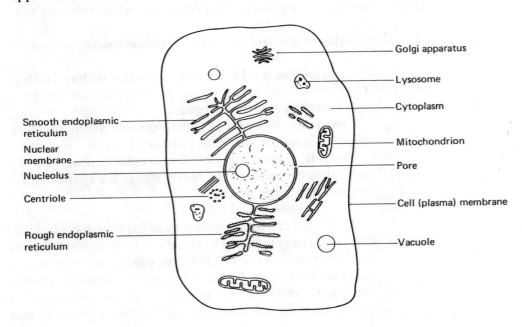

Smooth endoplasmic reticulum

Nuclear membrane

Nucleolus

Centriole

Rough endoplasmic reticulum

Golgi apparatus

Lysosome

Cytoplasm

Mitochondrion

Pore

Cell (plasma) membrane

Vacuole

Figure 1.1 *A typical animal cell*

These membrane-bound compartments are a characteristic feature of eucaryotic cells, and they include:

The nucleus   This is the most conspicuous compartment in the cell and it is bounded by a double membrane continuous with the endoplasmic reticulum; it contains the genetic material, the DNA, which together with associated acidic and basic proteins is referred to as **chromatin**. During cell division (Sections 1.3 and 1.4) the chromatin coils up and condenses, each DNA molecule and its associated proteins forming a **chromosome.**

The endoplasmic reticulum (ER)   This is a system of tubular membranes and flattened sac-like structures forming an intricate network of channels within the cytoplasm; it is continuous with the nuclear membrane. Some of the ER, the **rough ER,** has its outer surface lined with **ribosomes;** these complexes of RNA and protein are the cellular factories for protein synthesis. Other parts of the ER, the **smooth ER,** are devoid of ribosomes and play no part in protein synthesis.

The Golgi apparatus   This is a stack of flattened vesicles resembling the smooth ER. One of its functions appears to be to process and to package into vesicles (membranous sacs) secretory proteins made by the ribosomes lining the rough ER.

The mitochondria   Each mitochondrion consists of two layers of membrane with the inner layer forming extensive internal invaginations (**cristae**); mitochondria are rich in enzymes and they provide energy by the oxidation of food substances during oxidative respiration.

Lysosomes   These are vesicles formed in the Golgi apparatus and they contain enzymes involved in the breakdown (digestion) of a wide range of macromolecules.

Vacuoles   These small membrane-bound bodies are frequently used for storing food particles and other materials; they are not always present.

Other important but non-membranous components of animal cells are:

The ribosomes   The cellular factories within which protein synthesis occurs. They are mostly found lining the outer surface of the rough ER but many are also found free in the cytoplasm.

Centrioles   Systems of microtubules which organise the formation of the spindle during cell division (Section 1.3).

Nucleoli   These are globular structures observed in every nucleus and they mark the chromosomal locations where the ribosomal RNA molecules are being produced in large numbers.

Plant cells are rather different as (a) they have a thick rigid **cell wall** primarily composed of cellulose; (b) there is a large **central vacuole**, sometimes occupying the larger part of the cell volume, whose main function is to maintain the turgor pressure of the cell; (c) there are neither lysosomes nor centrioles; and (d) there are many **chloroplasts**, another type of organelle, which are involved in photosynthesis, the main energy-trapping process in plants.

1.2   Eucaryotic Chromosomes

Every nucleus contains a fixed number of linear bodies, the chromosomes, which carry the genetic information, and each chromosome consists of a single very long molecule of deoxyribonucleic acid (DNA) complexed with various proteins (Chapter 3). More specifically, the genetic information is encoded in the DNA, which is organised along its length into a large number of basic units called **genes**. These genes control all the metabolic activities of cells and mostly they function by specifying the production of particular polypeptides, often in the form of enzymes.

Eucaryotic cells usually contain two complete sets of chromosomes, one derived from each parent; the two members of each pair are referred to as **homologous chromosomes** or **homologues** and they carry identical sequences of genes. Thus, the genes also occur in pairs. However, the two representatives of any particular gene pair may be slightly different from each other and one may produce a gene product qualitatively or quantitatively different from that produced by the other; these alternative forms of the same gene are called **alleles**, and each is located at a specific position along one of the homologous chromosomes, the gene **locus**. If identical alleles are present at a particular locus on each member of a pair of homologous chromosomes, the cell or individual is said to be **homozygous** or a **homozygote**; if two different alleles are present, it is said to be **heterozygous** or a **heterozygote**.

1.3 Mitosis in an Animal Cell

When a cell divides, the continuity of the genetic material must be maintained and each daughter cell must receive an exact copy of each and every chromosome present in the parent cell. This is achieved by **mitosis**.

Interphase   When cells are not actively dividing, they are in the **resting stage** or **interphase;** the nuclei have a granular appearance, the chromosomes are not recognisable, as the DNA is fully extended, and one or more nucleoli are present (Figure 1.2.1). It is during interphase that the DNA replicates (Chapter 2).

Prophase   This marks the onset of cell division. The nucleoli break down, the chromosomes condense (i.e. become more tightly coiled) and become visible under the light microscope, and each is seen to be split longitudinally into two halves, the two **sister chromatids**, held together at an as yet undivided region, the **centromere** (Figure 1.2.2). As prophase proceeds, the centriole divides and the two daughters move to opposite ends of the nucleus, where they will form the poles of the mitotic spindle; the chromosomes condense still further and the centromere is clearly visible as a constriction (Figure 1.2.3).

Metaphase   A series of microtubules, the spindle fibres, grows outwards from each centriole; eventually these coalesce, to form the **mitotic spindle**, and each chromosome becomes attached to the mid-point of the spindle (the **equatorial** or **metaphase plate**) by its centromeres (Figure 1.2.4).

Anaphase   The centromeres now divide and the spindle fibres contract, so that the sister chromosomes (as they are known once the centromeres have divided) are drawn apart at their centromeres and move to opposite poles of the spindle. The result is that one copy of every chromosome is assembled at each of the poles (Figure 1.2.5).

Telophase   Cell division is completed by the re-formation of the nuclear membrane and the nucleoli, the uncoiling of the chromosomes and the formation

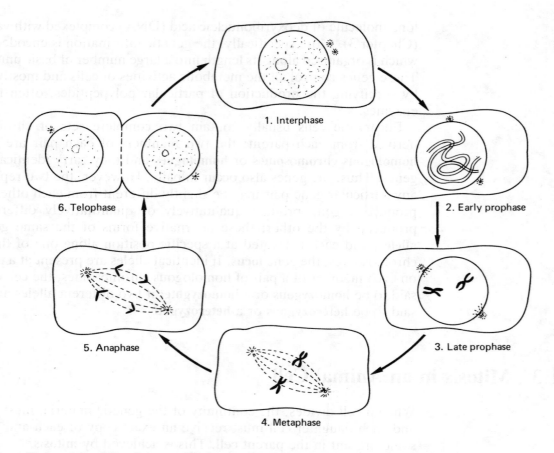

Figure 1.2 *Mitosis in a typical animal cell. The diagrams show the behaviour of one pair of homologous chromosomes*

of a new cell membrane between the daughter cells. At the end of telophase both nuclei are in interphase (Figure 1.2.6).

In higher plants mitosis is similar, but there are no centrioles and the spindle simply forms by the growth of microtubules between the two poles.

1.4 Meiosis

Higher eucaryotes, such as animals and plants, contain two complete sets of chromosomes (they are **diploid**) and, because of the accuracy of mitosis, every cell in an adult organism is genetically identical with the fertilised egg cell. This diploid **zygote** is formed by the fusion of a male and a female gamete, each containing only one set of chromosomes (the gametes are **haploid**), and the special type of cell division that halves the chromosome number during gamete formation is **meiosis** or **reduction division**. Thus, when the two gametes fuse to form a zygote, the diploid number is restored.

In meiosis there are *two* divisions of the cell but only *one* division of the chromosomes — thus, the chromosome number is halved.

(a) The First Meiotic Division

Prophase When the chromosomes first become visible, they are long and filamentous and, although the DNA has already replicated, they are not yet visibly divided into chromatids. They have a rather bumpy appearance, like beads on a string, and each bump is called a **chromomere**.

The pairs of homologous chromosomes now come together and pair, or **synapse**, along their length (Figure 1.3.1 and 2) and each pair of homologous chromosomes is referred to as a **bivalent**. The chromosomes now condense further and divide into two chromatids (Figure 1.3.3).

At this stage **chiasmata** are formed. Each chiasma has arisen as though two non-sister chromatids have broken at exactly homologous positions and the broken ends have reunited in a new combination (Figure 1.3.4); thus, genetic material has been reciprocally exchanged between non-sister chromatids. This is crossing-over, or recombination (see Chapter 5).

Crossing-over is an essential feature of meiosis; if, in a particular bivalent, no chiasma forms, then the two homologous chromosomes will separate and will be unable to align correctly on the metaphase plate.

Metaphase I The nuclear membrane breaks down and a spindle forms. Each bivalent attaches to the spindle by its centromeres, so that one centromere lies on each side of the metaphase plate (Figure 1.3.5).

Anaphase I The spindle fibres contract and the homologous centromeres are drawn to the opposite poles of the spindle (Figure 1.3.6). Note that the centromeres do not divide at anaphase I as they do in mitosis.

Telophase I The nuclear membranes re-form and the cell divides into two. This is the end of the first meiotic division and there may now be an extended interphase before the start of the second meiotic division.

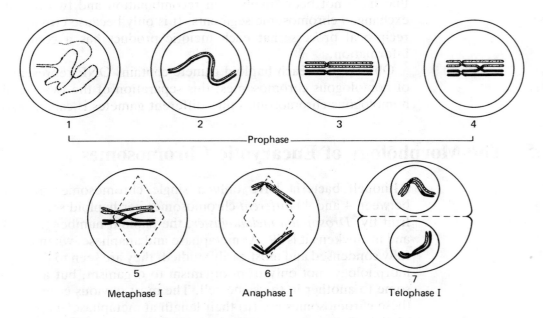

Figure 1.3 *The first meiotic division. The diagrams show the behaviour of one pair of homologous chromosomes. The small open circles represent the centromeres*

5

(b) The Second Meiotic Division

The second meiotic division takes place in each of the two products of the first meiotic division.

Metaphase II Two spindles form in the opposite plane to that of the first meiotic division and the chromosomes, already divided into chromatids, align on the metaphase plates of the newly formed spindles (Figure 1.4.1).

Anaphase II The centromeres finally divide and the contracting spindle fibres draw one daughter chromosome to each pole (Figure 1.4.2).

Telophase II Interphase nuclei are re-formed, resulting in the production of four haploid cells, each containing one complete set of chromosomes.

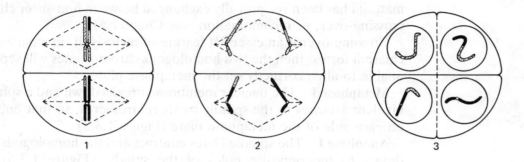

Figure 1.4 *The second meiotic division*

Each group of four cells resulting from one particular meiosis is a **tetrad**. Note that in each tetrad, for each recombination event, there are two chromosomes that have not been involved in recombination and two that carry reciprocally exchanged chromosome segments. It is only because crossing-over is an exactly reciprocal process that each meiotic product has a complete set of genetic information.

Observe that each haploid gamete contains ONE representative of each pair of homologous chromosomes; this separation of the two members of a pair of homologous chromosomes into different gametes is called **segregation**.

1.5 The Morphology of Eucaryotic Chromosomes

Although bacteria have only a single chromosome, most eucaryotes have between 4 and 40 *different* chromosomes per haploid set — for example, in the fruit fly, *Drosophila melanogaster*, the haploid number (n) is 4, in man it is 23 and in chickens it is 39. At metaphase and anaphase, when the chromosomes are fully condensed and most easily studied, they are seen to differ greatly in size and morphology, not only from organism to organism, but also from one chromosome to another in the same cell. The most obvious criteria used to distinguish these chromosomes are (a) their length at metaphase/anaphase; (b) the position of the centromere or **primary constriction**; (c) the position of the nucleolar organiser or **secondary constriction**, if present and visible.

At least one **nucleolar organiser** region is present in every haploid set of chromosomes, and this is the site where ribosomal RNA is synthesised and where a nucleolus will re-form during telophase; this region is relatively undercondensed and so appears as a constriction but it is not always visible.

The chromosome complement of a cell or organism is known as its **karyotype**, and this is frequently represented by drawing the metaphase or anaphase chromosomes and arranging them according to their length and the positions of their centromeres.

1.6   The Sex Chromosomes and Sex Determination

In many animals and some higher plants sex is determined by a special pair of chromosomes, called the **sex chromosomes** to distinguish them from all the other chromosomes, the **autosomes**. In *Drosophila melanogaster* there are three pairs of autosomes (chromosomes II, III and IV) and two sex chromosomes (Figure 1.5). In females the sex chromosomes are a pair of homologous rod-shaped **X chromosomes** but in males there are two dissimilar sex chromosomes, one X chromosome and a J-shaped **Y chromosome**; the females are designated XX and the males XY.

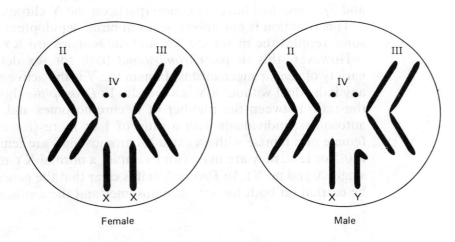

Female   Male

Figure 1.5   *The karyotypes of* D. melanogaster.
...omes are drawn as viewed from one of the poles of the spindle just after ... have divided — thus, the four pairs of chromosomes are arranged ...riphery of the metaphase plate with their arms hanging free in the cytoplasm.
...romosomes II and III are large and **metacentric** (they have median ...while chromosome IV is minute and dot-like. The X chromosomes are ...ized and **telocentric** (with a terminal centromere), while the Y, also ...niddle-sized, is **acrocentric** (the centromere is sub-terminal)

...les are the **homogametic** sex, since all the gametes they produce carry ...nosome, but the males are **heterogametic** and, as a result of meiosis, half their gametes are X-bearing and half are Y-bearing; thus, half the progeny of any mating will (on average) be XX females and half will be XY males:

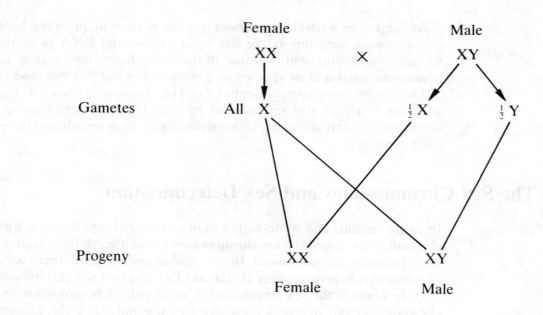

The X and Y chromosomes are generally non-homologous. Although the X chromosomes, like the autosomes, carry many genes of major effect (Chapters 4 and 5), these loci have no counterparts on the Y chromosome.

This situation is not universal, as in birds, lepidoptera, some amphibians and some reptiles the males are XX and the females are XY.

However, this simple *chromosomal* basis for sex determination conceals a variety of *genetic* mechanisms. In man the Y is the active arbiter of maleness and any individual without a Y is a female. In *Drosophila,* however, sex depends on the ratio between the number of X chromosomes and the number of sets of autosomes; individuals with a ratio of 1 or more (for example, a normal XX female or a female with an extra X chromosome) are females, while, if the ratio is 0.5 or less, they are male (for example, a normal XY male or a diploid with a single X and no Y). In *Drosophila* it is clear that the genes determining sex must be carried on both the sex chromosomes and the autosomes.

1.7 Procaryotic Cells

Procaryotes are defined by the absence of a nuclear membrane; they are much simpler organisms than eucaryotes and consist of two major groups, the bacteria and the blue-green algae (cyanobacteria). They are generally unicellular, very much smaller than eucaryotic cells and reproduce by simple binary fission, by producing spores or by budding.

Escherichia coli, the common gut bacterium, is a good example of a procaryote. It is important because it is one of the best-understood organisms at both a genetic and a molecular level; it is found in both man and animals as part of the bacterial flora of the intestine and, although it is normally harmless, some strains can cause diarrhoea.

E. coli (Figure 1.6) not only lacks a nuclear membrane, but also differs from eucaryotes in a number of other important respects:

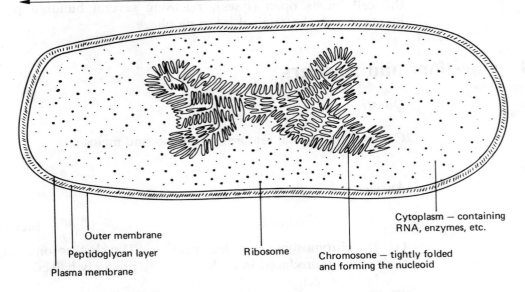

Figure 1.6 *Schematic representation of the rod-shaped cell of* E. coli

(1) Like plant cells, it has a cell wall outside the plasma membrane; however, its major component is not cellulose but peptidoglycan, a complex macromolecule formed by linking together sugar and polypeptide molecules. This layer is sandwiched between the plasma membrane and an outer membrane.

(2) No other permanent membranes are present, so there are no organelles such as mitochondria, chloroplasts and endoplasmic reticulum.

(3) The single chromosome is tightly folded into a rather diffuse region called the **nucleoid**.

(4) The chromosome has very little associated protein.

(5) It reproduces by binary fission. Mitosis does not occur and, after the DNA has replicated, one copy of the chromosome passes into each daughter cell.

(6) The ribosomes (which are smaller than in eucaryotes) exist freely in the cytoplasm.

1.8 Viruses

Much of our knowledge of molecular genetics has been obtained from studies using **viruses**, particularly the viruses that infect bacteria, known as **bacteriophages** or **phages**. Viruses consist of small pieces of genetic material (this is the viral chromosome and may be either DNA or RNA) packed into a protective protein coat, the **capsid**, and they are obligate parasites of animals, plants and bacteria. Since they lack the organelles required for protein synthesis, they are totally unable to reproduce or metabolise outside the host cell and they cannot be considered true living organisms.

When a typical phage infects a bacterial host cell, its chromosome replicates many times; these free chromosomes are then used to direct the synthesis of phage proteins using the synthetic machinery of the host cell. These proteins are

9

assembled into capsids, each containing a phage chromosome, and, eventually, the cell bursts open (**lyses**), releasing several hundred phage progeny, all identical with the original infecting phage.

1.9 Questions and Answers

Question 1.1

Contrast the special features of mitosis and meiosis.

Answer 1.1

Mitosis	Meiosis
(1) The chromosome complement is exactly reproduced in each daughter cell.	The chromosome number is precisely halved.
(2) There is one division of the chromosomes and one division of the cell.	There is one division of the chromosomes and TWO divisions of the cell.
(3) The chromosomes do not synapse.	Homologous chromosomes associate along their length (synapse) in pairs.
(4) There is no crossing-over.	Crossing-over occurs, resulting in chiasma formation and in genetic recombination.
(5) The chromosomes are visibly divided when they first appear at prophase.	The chromosomes are not *visibly* divided, although their DNA has already replicated.
(6) Does not reassort the existing genetic variation.	Reassorts the genetic variation, and results in the production of different phenotypes in the next generation.

Question 1.2

Write notes on (a) crossing-over and (b) chiasmata.

Answer 1.2

(a) **Crossing-over** occurs during early prophase of meiosis when homologous chromosomes are closely synapsed along their length. Each cross-over involves two non-sister chromatids and results in the reciprocal exchange of chromosome segments so that the two recombinant chromatids carry a new combination of the genes present on the two parental chromosomes (Figure 1.7).

Thus, crossing-over reassorts the genetic material, ensuring that genes on the same parental chromosome do not always segregate together at meiosis.

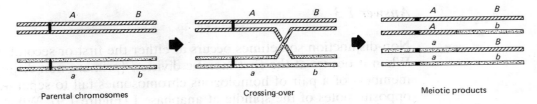

Parental chromosomes   Crossing-over   Meiotic products

Figure 1.7 *Crossing-over*

(b) Chiasmata, observed from mid-prophase to early anaphase, are the visible structures showing that crossing-over occurred at an earlier stage when the homologues were closely synapsed; they do NOT show the process of crossing-over.

The chiasmata hold the bivalent together and prevent the homologous chromosomes from separating; if there are no chiasmata, the homologous chromosomes separate and fail to orientate correctly on the metaphase plate. Thus, crossing-over and chiasmata are essential features of the meiotic process.

NOTES

1 The reassortment of the genetic material is explained in Section 5.1.

2 In procaryotic systems there is no equivalent to meiosis, but homologous DNA molecules can associate together and participate in crossing-over.

3 At anaphase I the homologous chromosomes move towards the opposite poles of the spindle and the chiasmata are slid towards the ends of the bivalent; this is **terminalisation** (Figure 1.8). This can be likened to pulling apart at the centre two strands of string that have been twisted together. The twists are slid towards the ends of the string and eventually disappear when the strands separate completely.

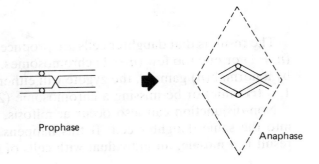

Prophase   Anaphase

Figure 1.8 *Terminalisation*

Question 1.3

What is non-disjunction and how does it give rise to cells with abnormal karyotypes?

11

Answer 1.3

Non-disjunction sometimes occurs at either the first or second meiotic division. When it occurs at the first meiotic division (primary non-disjunction), the two members of a pair of homologous chromosomes fail to separate and to pass to opposite poles of the spindle at anaphase I (Figure 1.9). When it occurs at the second meiotic division (secondary non-disjunction), two sister chromatids fail to separate at anaphase II. This is probably caused by the centromeres being incorrectly oriented on the metaphase plate.

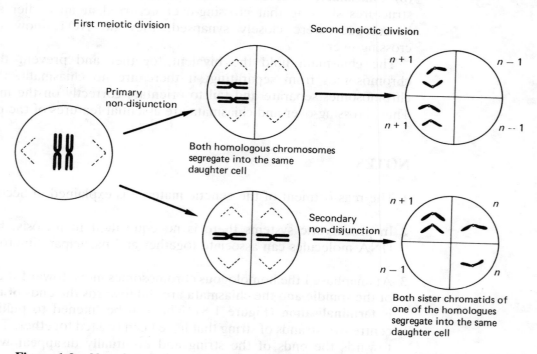

Figure 1.9 *Non-disjunction. Primary and secondary non-disjunction occur at the first and second meiotic divisions, respectively*

The result is that daughter cells are produced (i.e. gametes) with one too many ($n + 1$) or one too few ($n - 1$) chromosomes. When these gametes are fertilised by a normal (n) gamete, the zygote will either have an extra chromosome ($2n + 1$, a **trisomic**) or be missing a chromosome ($2n - 1$, a **monosomic**).

Non-disjunction can also occur at mitosis, when both sister chromatids pass into the same daughter cell. If this happens during somatic development, the result is a **mosaic**, an individual with cells of more than one genotype.

NOTE

Cells with extra or missing chromosomes are collectively referred to as **aneuploids**. In animals aneuploids usually have severe or lethal defects, but in plants many examples are known.

Question 1.4

Non-disjunction occasionally produces individuals who are either missing a chromosome or who have an extra chromosome. In *Drosophila* X0 flies (the zero shows that the second sex chromosome is missing) are sterile males but in humans X0 individuals are sterile females. Explain this difference.

What sex do you expect XXY individuals to be for both species?

Answer 1.4

In *Drosophila* sex is determined by the ratio between the number of X chromosomes and the number of sets of autosomes and the Y plays no apparent part. A ratio of unity (2X/2A) results in the development of a normal female and a ratio of 0.5 (1X/2A) a normal male. In man, however, the Y chromosome actively determines maleness, so that an X0 individual is a sterile female.

In *Drosophila* XXY flies have a ratio 2A/2X and are fertile females; XXY humans have a Y chromosome and are males (but they are sterile).

NOTES

1 In man aneuploidy for a sex chromosome does not produce gross morphological defects (as does aneuploidy for many of the autosomes), but it does result in abnormal sexual development and, usually, in a loss or reduction of fertility.

2 X0 females have **Turner's Syndrome**. The affected individuals have female external genitalia but their ovaries are only rudimentary; they are of short stature, have webbed necks and broad chests and a normal IQ. The syndrome affects about 1 in 3000 female births.

3 XXY males have **Kleinfelter's Syndrome**. They have male external genitalia but their testes are only rudimentary and do not produce sperm; they are tall and long-legged, they show some breast development and their IQ is usually within the normal range. The syndrome affects about 1 in 600 male births.

Question 1.5

Do you expect there to be greater genetic variation among the progeny of a sexually or an asexually reproducing organism?

Answer 1.5

In sexual reproduction meiosis is always involved in the formation of the gametes. During meiosis recombination occurs between the homologous pairs of paternal and maternal chromosomes and produces daughter chromosomes carrying new combinations of the genes present on the parental chromosomes;

furthermore, each chromosome segregates independently, so that different gametes contain different combinations of the paternal and maternal chromosomes. Thus, meiosis reassorts the existing genes so that there is considerable genetic and phenotypic variation in the next generation.

In eucaryotes asexual reproduction produces progeny by one or a series of *mitotic* divisions — for example, budding in yeast and vegetative propagation in plants. Thus, every daughter has exactly the same genetic constitution as its parent, and unless mutation occurs, there is no genetic variation.

NOTE

Meiosis reshuffles the genes but it does NOT produce any NEW genetic variation — this is produced only as a result of mutation (Chapter 11).

1.10 Supplementary Questions

1.1 Why is meiosis an essential part of the life cycle of a sexually reproducing organism?

1.2 Distinguish between a centriole and a centromere.

1.3 From your knowledge of the behaviour of the chromosomes at meiosis explain why the hybrid from a cross between two species is usually sterile.

1.4 The four drawings in Figure 1.10 illustrate the karyotypes of four individuals of *Drosophila melanogaster*. Karyotype A was obtained from a normal male; what can you infer from karyotypes B, C and D?

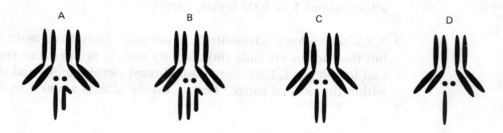

Figure 1.10

2 Nucleic acids

2.1 DNA and RNA

Before the 1940s it was generally thought that proteins were the genetic material, carrying and transmitting the genetic information required for development and reproduction; then, in 1944, Oswald T. Avery and his co-workers provided convincing evidence that **deoxyribonucleic acid (DNA)** was the active genetic principle. Nevertheless, many geneticists did not accept their conclusions and it was not until the early 1950s, by which time further evidence had accumulated, that it became accepted that DNA was the genetic material (Question 2.6). In all organisms the genetic information is stored, replicated and transmitted by DNA, the only exception being certain viruses, where another nucleic acid, **ribonucleic acid (RNA)**, carries the genetic information. However, as we shall see (Chapter 8), the principal functions of RNA are in protein synthesis.

2.2 The Structure of DNA

(a) The Building Blocks

Nucleic acids are assembled from building blocks called **nucleotides** into very-long-chain molecules known as **polynucleotides**; these vary in length from a few to many millions of nucleotides and include some of the largest known macromolecules. In turn, each nucleotide is assembled from three components (Figure 2.1):

(1) A pentose sugar molecule, **deoxyribose** in DNA and **ribose** in RNA; these are very similar but deoxyribose lacks the hydroxyl group (OH) present on the 2′ carbon atom of ribose.

(2) One or more **phosphate** groups.

(3) A nitrogenous base, which may be either a purine — **adenine** (A) or **guanine** (G) — or a pyrimidine — **cytosine** (C), **thymine** (T) or **uracil** (U, normally found only in RNA, where it replaces thymine). These bases are referred to as purines and pyrimidines because they are derived from the parental bases purine and pyrimidine (but these are not found in DNA and RNA).

In both DNA and RNA each base is joined to a sugar by a glycosylic bond between the 1′ carbon of the sugar and the nitrogen atom at position 9 of a purine or position 1 of a pyrimidine. This molecule is called a **nucleoside** — either a ribonucleoside or a deoxyribonucleoside. When one or more phosphate groups are added to the 5′ carbon of the sugar, the nucleoside is converted to the corresponding **nucleotide**. For example:

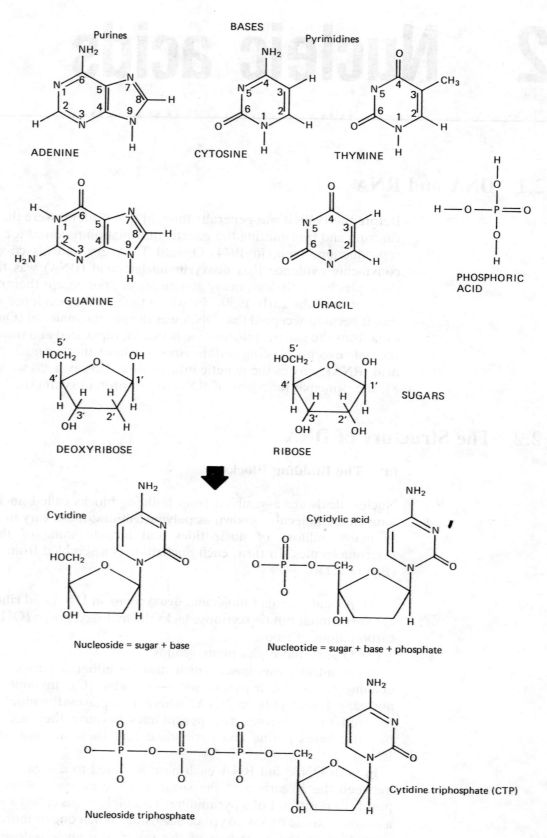

Figure 2.1 *The components of DNA and RNA*

	Base
Adenine	
A + deoxyribose	deoxyribonucleoside (known as deoxyadenosine)
A + deoxyribose + PO₄	deoxyribonucleoside monophosphate (deoxyadenylic acid, dAMP)
A + deoxyribose + PO₄ + PO₄	deoxyribonucleoside diphosphate (deoxyadenosine diphosphate, dADP)
A + deoxyribose + PO₄ + PO₄ + PO₄	deoxyribonucleoside triphosphate (deoxyadenosine triphosphate, dATP)

The other deoxyribonucleoside triphosphates are guanosine triphosphate (dGTP), cytidine triphosphate (dCTP) and thymidine triphosphate (dTTP); the corresponding ribonucleoside triphosphates are rATP, rGTP, rCTP and rUTP. These nucleoside triphosphates are particularly important, since they are the precursor molecules for nucleic acid synthesis.

(b) Polynucleotide Chains

In both DNA and RNA these nucleotides are joined together by **phosphodiester bonds** which connect the 3' position of one sugar with the 5' position of the next

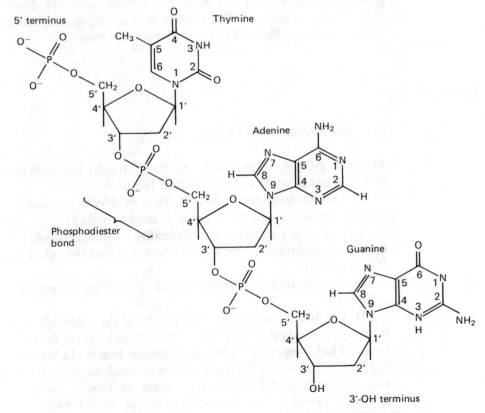

Figure 2.2 *The two-dimensional structure of a polynucleotide chain. Provided that a template strand is present (not shown), nucleoside triphosphates are added onto the 3'−OH end and two phosphate groups are released as pyrophosphate. This reaction is catalysed by DNA ligase*

sugar via a phosphate group (Figure 2.2). In effect there is a backbone of alternating sugar–phosphate groups with a base attached to each sugar:

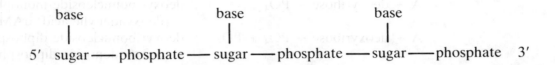

It is important to note that a polynucleotide has a **polarity**. In Figure 2.2 the top of the molecule has a free 5′ phosphate group, while the lower end has a free 3′ hydroxyl (—OH) group. The sequence of nucleotides shown can be read as 5′ pdT pdA pdG 3′ or 5′pT pA pG 3′ or, more simply, as TAG, remembering that nucleotide sequences are always written in the 5′ to 3′ direction.

When DNA is synthesised, new nucleotides are *always* added on at the 3′ end of the growing polynucleotide chain, so that synthesis only occurs in the 5′ to 3′ direction. Furthermore, only nucleoside triphosphates can participate in the formation of phosphodiester bonds and, since only one phosphate forms the actual bond, two phosphates are released each time a nucleotide is added.

Most RNA molecules are single-stranded (that is, in the form of a single polynucleotide chain) but most DNA molecules are double-stranded (Section 2.3). Single-stranded DNA only occurs as the genetic material of certain small viruses such as phage X174 and the parvoviruses.

2.3 DNA

(a) Double-stranded DNA

The biological structure of DNA, elucidated by James Watson and Francis Crick in 1953, has several very important features:

(1) Double-stranded DNA consists of TWO polynucleotide strands twisted around each other in the form of a **double helix** (Figure 2.3b). The two strands are of opposite polarity, or **antiparallel** — that is to say, one strand is oriented from top to bottom in the 5′ to 3′ direction and the other in the 3′ to 5′ direction (Figure 2.3a).

The double helix resembles a pair of ribbons wound around the outside of a cylinder.

(2) The bases project into the centre of the molecule and a base on one chain is always paired off with a base at the same level on the other chain: each pair of bases is held together by weak **hydrogen bonds** (H bonds).

(3) The pairing of the bases is not at random, as A is always paired with T by two H bonds and C with G by three H bonds: no other pairs of bases are normally possible. As a consequence, the two strands of the double helix have complementary sequences of nucleotides and they are said to be **base-complementary**.

(4) There is no restriction on the sequence of base pairs along the molecule and it is this sequence that encodes the genetic information (Chapter 9).

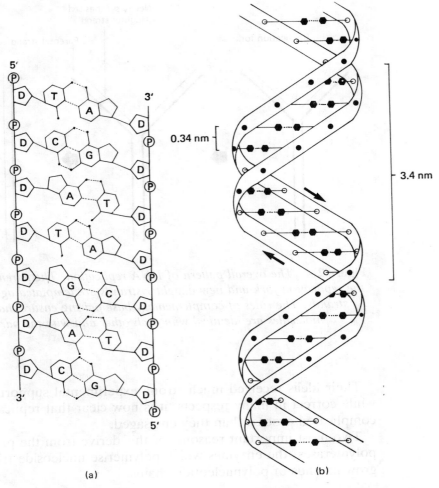

(a)   (b)

Figure 2.3   *The two- and three-dimensional structures of DNA.*
(a) The molecule consists of two polynucleotide chains, polarised in opposite directions and held together by weak H bonds (dashes) between complementary pairs of bases (A–T and C–G); D and P indicate the deoxyribose and phosphate groups, respectively.
(b) In three dimensions the two polynucleotide chains are in the form of a double helix. The ribbons represent the sugar–phosphate backbone chains. The bases and deoxyribose groups are denoted by the hexagons and circles, respectively

(b)   DNA Replication

In 1953 Watson and Crick proposed that DNA replicated by the two strands unwinding and separating so as to form a **replication fork**, each strand then acting as a **template** for the synthesis of a new daughter strand (Figure 2.4) by each exposed base attracting its complementary base in the form of a free nucleotide; once in place the free nucleotides could be joined together by the formation of phosphodiester bonds. This type of replication is called **semi-conservative**, since each daughter molecule consists of one parental (conserved) strand and one newly synthesised strand.

19

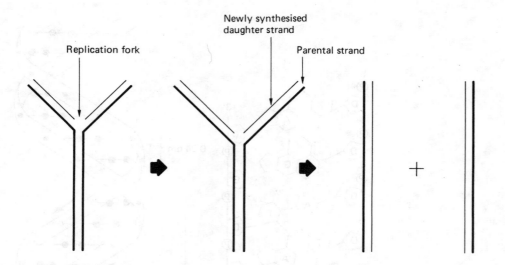

Figure 2.4 *The overall pattern of DNA replication. The parental duplex unwinds at the replication fork and new daughter strands are templated against the two parental strands. The rules of complementary base pairing ensure that the two daughter molecules are identical with each other and with the parental molecule*

Their ideas received much strong experimental support (Question 2.7) and, while correct in many respects, it is now clear that replication is a much more complicated process than they envisaged.

Two most important reasons for this derive from the properties of the DNA polymerases, the enzymes which polymerise nucleoside triphosphates onto the growing ends of polynucleotide chains:

(1) DNA polymerases (DP or DNAP) can ONLY add nucleotides onto a free 3'-OH end and NEVER onto the free 5' end; thus, only one new strand, the so-called **leading strand**, can be synthesised in the manner proposed by Watson and Crick. The other new strand, the **lagging strand**, is synthesised **discontinuously** as a number of small fragments, each about 1000 nucleotides long. Each fragment, known as an **Okazaki fragment**, is individually primed and polymerised in the 5' to 3' direction and then joined (ligated) to the preceding fragment (Figure 2.5 and Question 2.8). Although each fragment is synthesised in the 5' to 3' direction, the overall direction of growth is 3' to 5'.

(2) DNA polymerases can ONLY polymerise nucleotides onto an existing polynucleotide chain and so they can never initiate replication. Each newly synthesised tract of DNA (either a leading strand or an Okazaki fragment) must be **primed** by the synthesis of a short RNA primer onto which DNA polymerase can then add deoxyribonucleotides (Figure 2.5). This is possible because RNA polymerases, which join together ribonucleotides, do not require a primer.

The enzymes and other proteins involved in replication are assembled at the replication fork and as the complex moves along the molecule of duplex DNA it unwinds and replicates it. In *Escherichia coli*, where replication has been most extensively studied, the most important enzymes involved are:

20

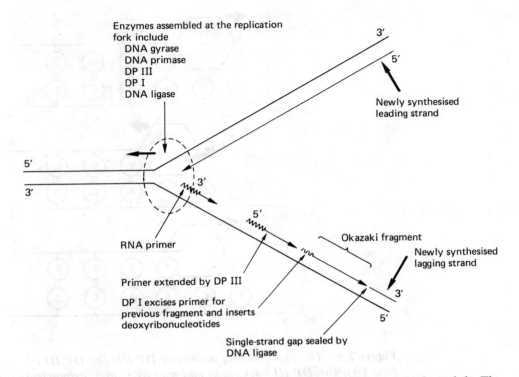

Enzymes assembled at the replication
fork include
  DNA gyrase
  DNA primase
  DP III
  DP I
  DNA ligase

3'

5'

Newly synthesised
leading strand

5'

3'

3'

RNA primer

5'

Okazaki fragment

Newly synthesised
lagging strand

Primer extended by DP III

3'

DP I excises primer for
previous fragment and inserts
deoxyribonucleotides

5'

Single-strand gap sealed by
DNA ligase

Figure 2.5 *DNA replication. The replication fork is moving from right to left. The newly synthesised upper strand is synthesised continuously, while the lower strand is discontinuously synthesised*

DNA gyrase Unwinds the double-stranded DNA.

DNA polymerase III (DP III) The normal replication enzyme. It polymerises nucleoside triphosphates onto the 3'—OH end of an existing polynucleotide chain (see Figure 2.2), using the parental strand as a template. This ensures that the new strand is exactly base complementary to the template strand. It also has 3' to 5' exonuclease activity — that is to say, it can digest either strand of duplex DNA from an exposed 3' end:

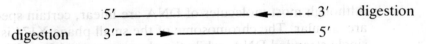

5'————————— 3' ◄ ─ ─ ─ ─ 3' digestion

digestion 3' ─ ─ ─► ————————— 5'

Thus, if DP III incorporates an incorrect nucleotide, it can back-track and use this exonuclease activity to excise the incorrect nucleotide; it then resumes its polymerase activity and inserts the correct nucleotide. This is **editing** or **proofreading** (Figure 2.6).

DNA polymerase I (DP I) This polymerises more slowly than DP III and has *both* 3' to 5' and 5' to 3' exonuclease activity:

(i) It recognises the exposed 5' ends of the RNA primers and uses its 5' to 3' exonuclease activity to digest them away and then uses its polymerase activity to replace them with the correct deoxyribonucleotides.

(ii) It follows DP III along the replicating DNA and can, using its 3' to 5' exonuclease and polymerase activities, correct any mistakes that have not already been corrected by DP III. Thus, it provides a fail-safe mechanism.

21

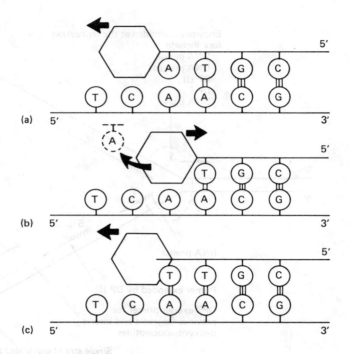

Figure 2.6 *The proofreading activity of DP III: (a) DP III has inserted an incorrect base (A); (b) DP III backtracks and uses its 3' to 5' exonuclease activity to excise the mismatched base; (c) DP III resumes its 5' to 3' polymerase activity and inserts the correct base (T)*

DNA primase This is a special RNA polymerase, and it primes the initiation of replication of the leading strand and of each Okazaki fragment by polymerising a short RNA chain (about 5 nucleotides long) against the DNA template.

DNA ligase Also known as polynucleotide ligase, this forms the phosphodiester bonds between two adjacent and unlinked nucleotides in a DNA duplex.

2.4 DNA Molecules May Be Circular

Although most molecules of DNA are linear, certain species of DNA molecule are circular. The chromosome of the small phage X174 is a circular molecule of single-stranded DNA, while the chromosome of *E. coli* is a circular molecule of double-stranded DNA (Chapter 3). The double-stranded DNA found in the mitochondria and chloroplasts of eucaryotes is also circular.

2.5 RNA

Chemically RNA differs from DNA in two ways; first, it contains ribose instead of deoxyribose; second, uracil replaces thymine (but note that uracil can pair with adenine in the same way as thymine). Otherwise it has the same structure as the single-stranded DNA molecule shown in Figure 2.2.

Although RNA is normally single-stranded, it is frequently folded into a complex stable secondary structure held together by H bonding between pairs of bases on the same strand. For example, if a strand of RNA contains two base-complementary sequences, one of which is in the reverse orientation to the other, it can form a hairpin by hydrogen bonding between the complementary bases with the intervening bases forming an unpaired loop:

```
                              A  C
                            G      U
CGAAUGCGACUGCAUGUA   ⟹      C≡G
                             G≡C
                             U=A
                             A=U
                           C G A    G U A
```

These sequences are known as **inverted repeats**. A single strand of RNA can also form duplex or partially duplex molecules by forming H bonds with a base-complementary or partially base-complementary strand of either RNA (forming an RNA–RNA duplex molecule) or DNA (forming an RNA–DNA hybrid duplex).

The most important types of RNA are:

Chromosomal RNA This is called informational RNA because it stores the genetic information. RNA is the genetic material of some viruses, including phage MS2 and the tobacco mosaic, influenza and polio viruses. In one exceptional instance, in a virus-like particle found in some yeasts, this RNA is double-stranded.

Messenger RNA, mRNA.
Ribosomal RNA, rRNA.
Transfer RNA, tRNA.

These RNAs are synthesised by the enzyme **RNA polymerase** (RP or RNAP), which polymerises ribonucleotides, using one of the strands of a duplex DNA molecule as a template: this process is called **transcription** (Chapter 8). Thus, these RNAs are all base complementary to one of the strands of DNA and have the same base sequence as the non-transcribed DNA strand (with uracil replacing thymine).

2.6 Questions and Answers

Question 2.1

Nucleic acids isolated from four different species had the following base ratios (%):

	A	T	U	G	C	$\dfrac{A+T \text{ (or } A+U)}{G+C}$	$\dfrac{A+G}{C+T \text{ (or } C+U)}$		
species 1	17	17		33	33	0.5	1.0	DNA	D·S
2	29	19		22	30	0.97	1.0	DNA	S·S
3	24		16	24	36	0.66	1.5	RNA	
4	3	34	16	16		2.1	1.0	RNA	D·S

23

(a) For each species state whether

(i) the nucleic acid is DNA or RNA;

(ii) it is single-stranded or double-stranded.

(b) Complete the missing percentages for species 4.

Answer 2.1

(a) Species 1 has DNA (since T present and U absent), and, since A = T and C = G, this DNA is double-stranded.

Species 2 also has DNA but, since A ≠ T and C ≠ G, this is single-stranded. (Note that both the A+T/G+C and A+G/C+T ratios are unity, although this DNA is single-stranded.)

Species 3 has RNA (since U present and T absent) and this is single-stranded.

Species 4 has double-stranded DNA (since T is present and A+G/C+T = 1).

(b) The missing percentages are A, 34; C, 16; and G, 16.

Question 2.2

The molecular weight of the *Escherichia coli* chromosome is about 2.5×10^9 daltons, the average weight of a nucleotide is 330 daltons and the distance between two adjacent nucleotide pairs is 0.34 nm; the DNA double helix makes one complete turn every 3.4 nm.

(a) How long is the molecule?

(b) How many turns does the DNA contain?

Answer 2.2

(a) 1 base = 330 d. Thus, 1 base pair (bp) = 660 d.

Number of base pairs = $2.5 \times 10^9/660 = 3.8 \times 10^6 = 3800$ kb (kilobases)

Each base pair is 0.34 nm apart. Thus,

length of chromosome = $3.8 \times 10^6 \times 0.34$ nm
$$= 1.3 \times 10^6 \text{ nm}$$
$$= 1.3 \text{ mm}$$

(b) Number of turns in double helix = $3.8 \times 10^6 \times 0.34/3.4 = 3.8 \times 10^5$.

Question 2.3

List the differences in the structure of bacterial DNA and RNA.

Answer 2.3

DNA	RNA
Thymine present	Uracil replaces thymine
Sugar is deoxyribose	Sugar is ribose
Double-stranded	Single-stranded
A+G/C+T is unity	A+G/C+T is not unity
Circular molecule	Linear molecule

Question 2.4

At one time it was thought that DNA, irrespective of its source, was a regularly repeating array of the four nucleotides (ATCG.ATCG.ATCG.ATCG. . . for example) and so lacked the specificity to be the genetic material. What was the first evidence that directly disproved this tetranucleotide theory?

Answer 2.4

Between 1949 and 1951 Erwin Chargaff found that:

(1) The base composition of DNA varies widely from source to source.
(2) There were nearly always equivalent amounts of A and T and of C and G (this is known as Chargaff's rule).
(3) Although the ratio A+G/C+T was always unity, the ratio A+T/G+C varied widely from organism to organism.

Question 2.5

Why are adenine–thymine and cytosine–guanine the only base pairs normally found in DNA?

Answer 2.5

(1) A C–G base pair is too large to be accommodated within the double helix, whereas an A–T base pair is too small and the gap between the nucleotides is too large for H bonds to form.
(2) A and T normally have two H bonds, while C and G have three. A base with two bonds cannot normally pair with a base with three bonds.

Question 2.6

List some of the evidence that first demonstrated that DNA (or RNA), and not protein, was the genetic material.

Answer 2.6

(1) In the 1940s Oswald Avery and his co-workers showed that DNA extracted from a virulent smooth (enclosed in a polysaccharide capsule) strain of the bacterium *Streptococcus pneumoniae* was able to be adsorbed by an avirulent and rough (lacking a capsule) strain, converting, or **transforming**, a few of these cells to the virulent smooth type. This transformation did NOT occur if the extracted DNA was first treated with DNase to degrade the DNA.

(2) In 1956 Heinz Fraenkel-Conrat reconstituted tobacco mosaic viruses (TMV), making a hybrid with the protein coat of one strain and the RNA of another strain. (*Note:* The genetic material of TMV is a single-stranded molecule of RNA.) When these hybrids infected tobacco plants, then (i) the lesions produced were always typical of the strain from which the RNA was obtained, and (ii) the progeny viruses isolated from these lesions had both the RNA and the protein coats characteristic of the parental strain providing the RNA.

(3) When a phage particle infects a bacterial cell, ONLY the phage nucleic acid enters the infected cell (although there may sometimes be a minute amount of associated protein as well) and this is sufficient to encode the production of complete new phage particles.

(4) Chemicals which specifically alter the structure of DNA can induce heritable changes or **mutations**.

(5) In any species the amount of DNA is constant from cell to cell, except that the haploid gametes contain exactly one-half of this amount. This is expected if DNA is the genetic material. Furthermore, this constancy and relationship is not found for any other cellular component.

Question 2.7

Describe the experiment carried out by Matthew Meselson and Franklin Stahl to demonstrate that DNA replicates semiconservatively.

Answer 2.7

(1) They grew cells of *E. coli* in ^{15}N medium for several generations, so that the DNA was labelled 'heavy' in both strands.

(2) The cells were transferred to ^{14}N medium, so that all the new strands of DNA were labelled 'light'.

(3) At various times the cells were removed from the ^{14}N medium and the DNA was extracted.

(4) The DNA was dissolved in caesium chloride and centrifuged at 100 000 g. This was the then newly developed technique of density gradient centrifugation.

(5) After many hours equilibrium was reached, when the forces of diffusion exactly balanced out the forces of centrifugation. Thus, the DNA concentrates in bands in the centrifuge tube, each band having the same density as the caesium chloride at that point.

(6) The position of each band, and the amount of DNA it contained, were determined photographically (Figure 2.7).

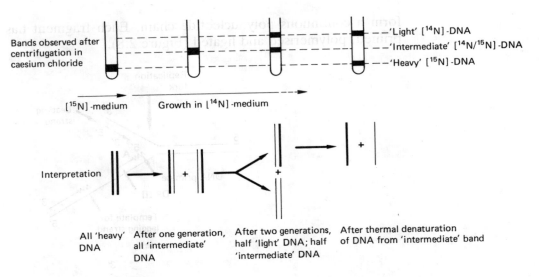

Bands observed after centrifugation in caesium chloride

'Light' [¹⁴N]-DNA
'Intermediate' [¹⁴N/¹⁵N]-DNA
'Heavy' [¹⁵N]-DNA

[¹⁵N]-medium   Growth in [¹⁴N]-medium

Interpretation

All 'heavy' DNA | After one generation, all 'intermediate' DNA | After two generations, half 'light' DNA; half 'intermediate' DNA | After thermal denaturation of DNA from 'intermediate' band

Figure 2.7 *The Meselson and Stahl experiment*

They found

(i) After growth in the heavy medium all the DNA concentrated in a single heavy band.

(ii) After a time corresponding to one generation cycle in ¹⁴N medium, all the DNA concentrated in a band of intermediate density, between the bands formed by pure ¹⁴N and pure ¹⁵N DNA. This is expected if replication is semiconservative, since each daughter molecule will have one old heavy and one new light strand.

(iii) After a further generation in ¹⁴N, one-half of the DNA was of intermediate density and one-half was light — again as expected with semiconservative replication. In further samples the proportion of intermediate density DNA decreased as expected.

(iv) Finally, they convincingly demonstrated that the first generation molecules are two-stranded and must replicate semiconservatively. This they achieved by heating the 'hybrid' DNA from the first generation so that the molecules **denatured** into their two component strands; when this DNA was recentrifuged, they found one heavy and one light band.

Question 2.8

Explain how the lagging strand is synthesised during DNA replication.

Answer 2.8

Since DNA polymerases only synthesise in the 5′ to 3′ direction, the lagging strand cannot be made in the same overall direction as the leading strand. The lagging strand is synthesised as a number of separately made fragments (**Okazaki fragments**), each made in the 5′ to 3′ direction; these are then ligated together to form a continuous polynucleotide chain. Each fragment has to be separately primed, polymerised and ligated (Figure 2.8).

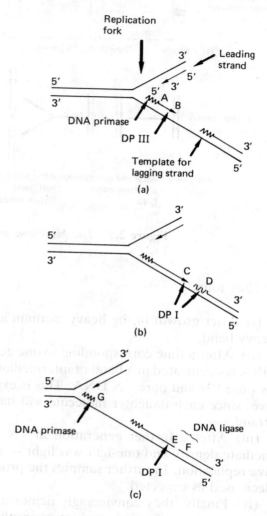

Figure 2.8 *Discontinuous synthesis of the lagging strand.*
(a) DNA primase recognises a short priming sequence on the template strand and synthesises a short RNA primer (A): DP III polymerises nucleotides on to the 3' end of the primer (B).
(b) DP III continues to add on nucleotides (C) until it reaches the primer for the previously synthesised fragment (D).
(c) DP I recognises the free 5' end of the primer and uses its exonuclease activity to digest it away. Simultaneously it uses its polymerase activity to fill in the gap with the corresponding deoxyribonucleotides (E). This leaves a gap in the sugar–phosphate backbone which is sealed by DNA ligase making a phosphodiester bond between the 3' end of (E) and the 5' end of the previous fragment (F). Meanwhile the replication fork has moved on and DNA primase has made another primer at the next priming sequence (G).

Question 2.9

Describe rolling-circle replication and list its special features.

Answer 2.9

Only certain circular molecules of DNA can replicate by the rolling-circle mode and the process is involved in the vegetative replication of certain phages (including λ) and in the transfer replication of the conjugal plasmids (see Chapter 6). Rolling-circle replication occurs as shown in Figure 2.9.

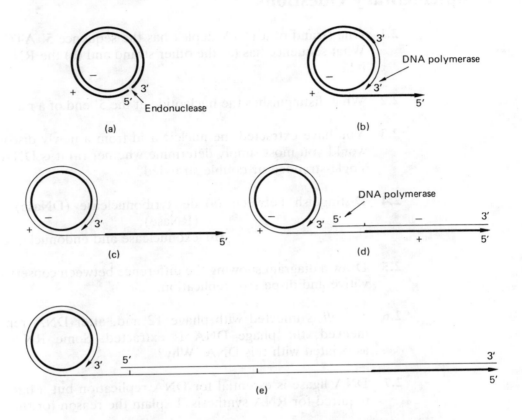

Figure 2.9 *Rolling-circle replication of DNA*
(a) One strand of a circular molecule (the (+) strand) is nicked by endonuclease.
(b) The 5' end is peeled off as a single strand and a new (+) strand is synthesised, using the intact (−) strand as a template.
(c) As the (+) strand is peeled off, it is continuously elongated at the 3' end.
(d) The tail may now become double-stranded by a new (−) strand being synthesised against the peeled-off (+) strand.
(e) The usual result of rolling-circle replication is a concatamer, an end-to-end array of a number of unit length genomes

The special features of this process are:

(1) Although replication is semiconservative, it is unidirectional and asymmetric.

(2) The immediate product is single-stranded DNA but this may be converted to double-stranded DNA by the synthesis of a complementary strand.

(3) The daughter molecules may be concatenated — that is, an end-to-end array of identical DNA molecules each corresponding to one unit genome.

(4) The concatenated DNA can subsequently be cut into pieces each corresponding to a single genome.

(5) The (−) strand always remains circular and so maintains intact one complete set of genetic information.

2.7 Supplementary Questions

2.1 One strand of a DNA duplex has the sequence 5′ ATCATGCCAGG 3′. What sequence has (a) the other strand and (b) the RNA transcribed from it?

2.2 What distinguishes the nucleotide at the 5′ end of a nucleic acid molecule?

2.3 You have extracted the nucleic acid from a newly discovered virus. How would you most simply determine whether (a) it is DNA or RNA and (b) single-stranded or double-stranded?

2.4 Distinguish between (a) deoxyribonuclease (DNase) and ribonuclease (RNase);
(b) exonuclease and endonuclease.

2.5 Draw a diagram showing the difference between conservative, semiconservative and dispersive replication.

2.6 *E. coli* is infected with phage T2 and, after DNA replication has commenced, the phage DNA is extracted. Some RNA is found closely associated with this DNA. Why?

2.7 DNA ligase is essential for DNA replication but a ligase is not normally required for RNA synthesis. Explain the reason for this.

2.8 At one time it was suggested that DNA replicated conservatively, each duplex molecule acting as a template for a new daughter duplex. If this were so, what results would Meselson and Stahl have obtained in their experiments?

3 Chromosome organisation

3.1 The Chromosomes of Viruses and Bacteria

The simplest chromosomes, the structures that carry the genetic information and transmit it from one generation to the next, are those of the viruses, as they consist of a single molecule of either DNA or RNA with no associated chromosomal proteins. The following examples illustrate the range of viral chromosomes:

(1) **Tobacco mosaic virus (TMV)** has a linear molecule of single-stranded RNA only 6400 nucleotides long.

(2) **Coliphage T2** (coliphages infect *E. coli*) is a large phage and has a linear molecule of double-stranded DNA about 165 kb long.

(3) **Coliphage lambda** (λ) has a linear molecule of double-stranded DNA 48.6 kb long, but as soon as this molecule enters the infected cell, it circularises; this is possible because the opposite ends of the linear molecule of DNA are single-stranded and base-complementary — thus, these **'sticky' ends** can associate by H bonding between the complementary bases and DNA ligase can seal the remaining single-strand gap (Question 3.7).

(4) **Coliphage X174** is a very small polyhedral phage and its chromosome is a circular molecule of single-stranded DNA, only 5.4 kb long and encoding 11 proteins. After infection this molecule becomes double-stranded by the synthesis of a complementary strand and then enters a complex replication cycle which results in the production of many single-stranded circular molecules of phage DNA.

The molecules of both lambda and X174 DNA are small enough to be visualised by electron microscopy.

Bacteria have very much larger chromosomes. The chromosome of *E. coli* is circular and consists of 3800 kb of double-stranded DNA folded into about 100 loops, each containing about 40 kb of DNA (Figure 3.1). In turn, each loop consists of 160–180 bead-like structures, each containing 220–265 bp of DNA. Four proteins are associated with this structure and are probably responsible for holding it together.

Although too large to be seen by electron microscopy, replicating chromosomes of *E. coli* can be visualised under the light microscope by using the

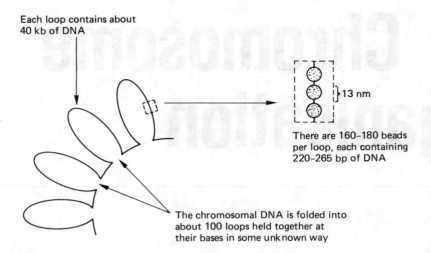

Each loop contains about
40 kb of DNA

13 nm

There are 160–180 beads
per loop, each containing
220–265 bp of DNA

The chromosomal DNA is folded into
about 100 loops held together at
their bases in some unknown way

Figure 3.1 *Chromosome organisation in* E. coli. *The DNA is organised into a diffuse structure called the nucleoid*

autoradiographic technique first developed by John Cairns in 1963 (Question 3.1).

The chromosome of *E. coli*, like most circular molecules of DNA, replicates from a single replication point, the **origin**, but in *both* directions (Figure 3.2). This is known as **theta-** (θ−) or **Cairns-type** replication, and the process is terminated when the two replication forks meet at the opposite side of the molecule: the two daughter molecules are released and, when the bacterium divides, one chromosome passes into each daughter cell.

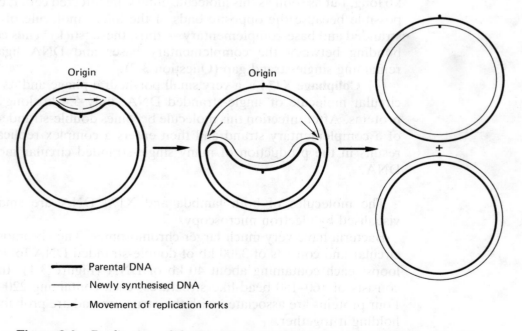

Origin   Origin

—— Parental DNA

—— Newly synthesised DNA

→ Movement of replication forks

Figure 3.2 *Replication of the* E. coli *chromosome. Replication commences from a fixed origin (*origin*) and proceeds bidirectionally*

3.2   The Eucaryotic Chromosome

(a)   Molecular Organisation

Each eucaryotic chromosome contains a single very long molecule of DNA (an average mammalian chromosome contains about 5 cm of DNA) and, in order to be accommodated within the nucleus, it has to be folded in a very organised way. This is possible because the DNA is complexed with five highly charged basic proteins known as **histones H1, H2A, H2B, H3 and H4;** this nucleohistone complex maintains the structural integrity of the chromosome. In addition, the chromosome contains some RNA, probably newly transcribed RNA, and many different non-histone proteins in variable quantities; the latter are probably involved in transcription (e.g. RNA polymerase) and gene regulation, and they do not appear to be an integral part of the chromosome. The entire nucleo-protein complex is referred to as **chromatin**.

At the first level of organisation, two turns of a continuous molecule of DNA are spooled around successive octamers of histone protein: each octamer is made up from two molecules each of histones H2A, H2B, H3 and H4, and each successive nucleohistone complex plus the intervening **linker** DNA forms a **nucleosome** (Question 3.1). Each nucleosome contains 150–270 bp of DNA. This organisation results in a fibre 10 nm in diameter, and this is now wound in the form of a solenoid to form a 30 nm fibre. This is the form of DNA in interphase cells, but at cell division further levels of coiling occur and, overall, the DNA is compacted over 5000-fold and the entire nucleoprotein structure is visible by light microscopy as a **metaphase chromosome** (Figure 3.4; Question 3.1).

(b)   The Replication of Eucaryotic DNA

In eucaryotic cells the DNA replicates during the preceding interphase and before the chromosomes commence to condense. In contrast to the DNA of viruses and bacteria, which replicate from a single replicational origin, most eucaryotic DNA commences to replicate from *many* separate initiation sites; each individually replicated segment is about 20–30 μm long (about 60–90 kb) and replicates bidirectionally (Figure 3.3). As the DNA replicates, so it becomes complexed with histones and organised into nucleosomes.

(c)   Types of Chromatin

Two types of chromatin are found in most eucaryotic chromosomes:

(1) **Euchromatin**   In interphase cells this probably exists as extended 30 nm fibres, almost invisible under the light microscope, but as cell division approaches, it becomes highly condensed and deeply staining. It is the major type of chromatin, it is genetically active and it contains the larger part of the genes. These genes are only active during interphase, when the euchromatin is fully extended: when chromatin is condensed, the genes cannot be transcribed and so remain inactive.

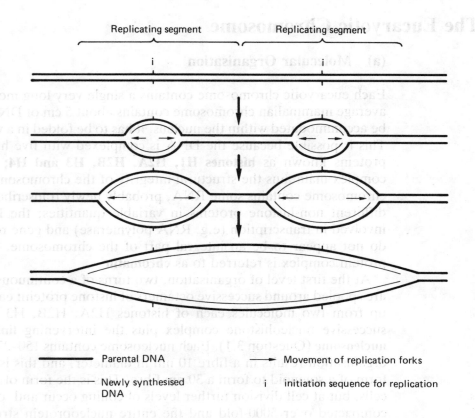

Figure 3.3 *The replication of eucaryotic DNA. Replication commences at many separate initiation points. Each replicated segment consists of 20–30 μm of fully extended DNA*

(2) Heterochromatin   This does not display the regular cycle of extension and condensation which is shown by euchromatin. There are two types of heterochromatin: **constitutive heterochromatin** and **facultative heterochromatin.**

Constitutive heterochromatin is relatively highly condensed and darkly staining at all stages of the cell division cycle (but it is less condensed than is euchromatin during cell division). Constitutive heterochromatin is a *permanent* part of the genome and, because it is always condensed, it can never be transcribed; it is probably genetically inert and it does not appear to contain any structural genes (i.e. genes encoding RNA and proteins) and it may be involved in maintaining the structural integrity of the chromosome. It also replicates after the euchromatin.

Constitutive heterochromatin is associated, in particular, with the centromeric regions of chromosomes, but it also occurs as 'blocks' distributed throughout the chromosome complement of most eucaryotes. Because it is a permanent feature, constitutive heterochromatin is always present at homologous sites on *both* members of each pair of chromosomes.

Facultative heterochromatin has the potential to become heterochromatic, but it only does so at certain periods in certain cells — at all other times it is fully extended, like euchromatin. It does contain structural genes, but these genes are only active when the chromatin is in the euchromatic state. Thus, facultative heterochromatinisation is a mechanism for turning off the activity of whole chromosomes or of large blocks of genes on particular chromosomes. An excellent example of facultative heterochromatin is the inactive X chromosome found in female mammals (Question 3.5).

In Section 1.5 we listed the principal features which distinguish one chromosome from another. To this we can now add the position of any blocks of heterochromatin, as revealed by staining methods.

(d) The Centromere

After the DNA has replicated and the chromosomes have condensed, each metaphase chromosome consists of two identical sister chromatids (Figure 3.4) held together only at an as yet undivided region of the chromosome, the **centromere**.

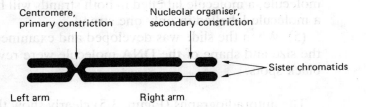

Figure 3.4 labels: Centromere, primary constriction; Nucleolar organiser, secondary constriction; Sister chromatids; Left arm; Right arm

Figure 3.4 *Outline drawing of a typical eucaryotic chromosome at metaphase. The heterochromatin of the centromere appears as a constriction; this heterochromatin is particularly late replicating, so that the sister chromatids cannot separate until anaphase*

The centromere is a highly specialised region of the chromosome lying within a block of constitutive heterochromatin and, in metaphase chromosomes, it appears as a constriction, the **primary constriction**. It has two very important special properties:

(1) It contains a short core sequence (in yeast this is 220 bp), probably associated with a special non-histone protein which binds the chromosome to the spindle fibres in the region of the metaphase plate.

(2) It does not replicate until the start of anaphase (remember that heterochromatin is late replicating); thus, as the spindle fibres contract, one sister chromatid (now called a daughter or sister chromosome) is drawn to each pole of the spindle. Every chromosome behaves in this way, so that one set of daughter chromosomes is assembled at each pole of the spindle.

3.3 Questions and Answers

Question 3.1

Describe the autoradiographic technique used by John Cairns to visualise replicating chromosomes of *E. coli*.

Answer 3.1

In 1963 he developed an autoradiographic technique which enabled him to visualise replicating chromosomes under the light microscope.

(1) *E. coli* was grown for up to two generations in tritiated ($[^3H]$) thymidine, so that any newly synthesised DNA was labelled 'hot'.

(2) The cells were very gently broken open with lysozyme, and the DNA was extracted and allowed to settle on a dialysis membrane.

(3) The membrane was mounted on a microscope slide, covered with a photographic emulsion and stored in the dark for 2 months.

(4) During storage the tritium particles decay and emit beta particles; these expose the photographic emulsion and, since they only travel about 1 μm, the position of the silver grains in the emulsion indicates the actual location of the tritium label in the chromosome. Further, the number of silver spots on the emulsion measures the density of the tritium label in the underlying DNA molecule; a molecule labelled in both strands will have twice the grain density of a molecule labelled in only one strand.

(5) When the slide was developed and examined under the light microscope, the size and shape of the DNA molecule were revealed by a continuous line of black spots.

The autoradiographs (Figure 3.5) clearly show that the molecule is circular, is about 1100 μm in circumference, and replicates semiconservatively. The molecule represented in the figure commenced to replicate at exactly the time the tritium was introduced, but the examination of molecules with more complex patterns of labelling showed that successive cycles of replication commenced from the same replication origin. Cairns also concluded that the molecule only replicated in one direction, but we now know that replication is normally bidirectional.

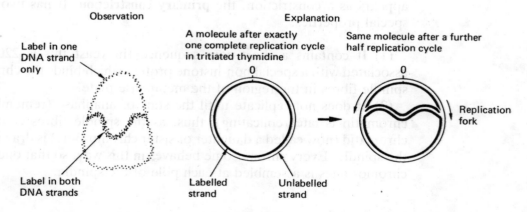

Figure 3.5 *Autoradiographic visualisation of the* E. coli *chromosome*

Question 3.2

Show how J. H. Taylor demonstrated that each eucaryotic chromosome behaves as if it were a single very long molecule of DNA.

Answer 3.2

In 1957 J. Herbert Taylor carried out a Meselson and Stahl type of experiment on *whole chromosomes* rather than on *molecules of DNA*.

(1) Seedlings of the broad bean, *Vicia faba*, were grown in tritiated ($[^3H]$) thymidime, so that any newly synthesised DNA was labelled 'hot'.

(2) After 8 h, when all the DNA was fully labelled, the roots were transferred to normal medium containing colchicine. The latter prevented spindle formation, so that the chromosomes divided but not the cells. Thus, daughter chromosomes were retained in the same cell and the total number of chromosomes present indicated the number of replication cycles that had occurred.

(3) Root tips were squashed onto microscope slides and the chromosomes were stained in the usual way; the slides were then covered with a photographic emulsion sensitive to the beta particles emitted by the tritium.

(4) The slides were stored in the dark. During this time each beta particle emitted exposed a spot on the emulsion.

(5) The photographic emulsions were developed *in situ*. Thus, the pattern of spots, showing the distribution of the isotope, could be compared with the actual underlying chromosomes.

Results

(i) After growth in tritiated thymidine, both chromatids of each metaphase chromosome were labelled (Figure 3.6).

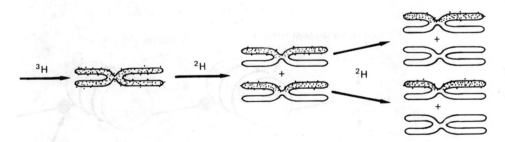

Figure 3.6 *Taylor's experiments with* Vicia faba

(ii) After a further generation in normal medium, one chromatid was labelled and the other was unlabelled.

(iii) After two generations, one-half of the chromosomes were unlabelled and one-half had one chromatid labelled and one unlabelled.

Conclusion

That each chromosome is a duplex structure, one part of which is passed into each sister chromatid at the time of chromosome replication. Thus, the chromosome replicates semiconservatively and behaves here as if it were a single very long molecule of DNA.

Further information

This experiment was important because at that time very little was known about chromosome structure and there was no evidence to suggest that each chromosome contained only one molecule of DNA.

Question 3.3

Describe the molecular organisation of a eucaryotic chromosome.

Answer 3.3

The structural components of interphase (non-dividing) chromosomes are (1) DNA, the genetic material, and (2) five-histone proteins. When chromatin from interphase cells is examined by electron microscopy two types of fibre are seen: (i) bumpy fibres 10 nm in diameter and (ii) smoother 30 nm fibres.

In the 10 nm fibres the DNA is at the first order of organisation and each bead, or **nucleosome**, consists of 146 bp of DNA spooled around an octamer of two molecules each of four histones (H2A, H2B, H3 and H4). The DNA is continuous from one nucleosome to the next and two turns are made around each octamer. The fifth histone (H1) is associated with the linker DNA connecting adjacent nucleosomes. This organisation is shown in Figure 3.7 and compacts the DNA by a factor of 5. Note that the DNA is spooled around the outside of the histone octamer and so remains available for gene expression.

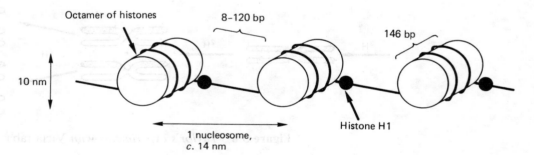

Figure 3.7 *The organisation of DNA and histones in chromosomes*

The 10 nm fibre is now further condensed by being wound into a solenoid with six nucleosomes per turn; this produces the 30 nm fibre in which the DNA is compacted 40-fold. It appears that histone H1 is essential for the maintenance of this second order of organisation.

These fibres are too fine to be seen under the light microscope, but, when cells are dividing, the metaphase chromosomes are clearly visible and the chromatin is obviously further compacted. This is achieved by the chromatin becoming organised into a series of large loops, each containing 35–100 kb of DNA, which are attached at their bases to a dense network of non-histone protein fibres called the **metaphase scaffold**. This scaffold is the same size and shape as the metaphase chromosome. How it is assembled is unknown, but in the metaphase chromosome the DNA is compacted some 5000-fold. For example:

An average human chromosome contains 50 000 μm (5 cm) of DNA.
The 10 nm fibre (compaction × 5) would be 10 000 μm.
The 30 nm fibre (compaction × 40) would be 1250 μm.
Since the average metaphase chromosome is 10 μm long, the final degree of compaction is 5000.

Question 3.4

(a) What is a chromomere?

(b) Describe a polytene chromosome.

(c) Indicate the special use of polytene chromosomes.

Answer 3.4

(a) When chromosomes first become visible at meiotic prophase, they have the appearance of beads on a string; in the beads, or **chromomeres**, the DNA is more condensed than in the interchromeric regions. The bands seen along polytene chromosomes (below) are also termed chromomeres, but there is no reason to suppose that these correspond to the chromomeres seen along meiotic chromosomes.

(b) In certain tissues of dipteran larvae, notably the salivary glands, the cells continue to grow in size but cell division is arrested. In each cell the two members of each homologous pair of chromosomes pair (synapse) along their length, as they do at meiotic prophase, and repeatedly divide but without separation of the daughter strands. The result is a multistranded or **polytene** chromosome consisting of 1024, 2048 or even 4096 (11 successive replications) individual strands of chromatin.

$$2 \longrightarrow 4 \longrightarrow 8 \longrightarrow 16 \longrightarrow 32 \; \text{---} \; \text{up to 4096}$$

The two original chromosomes, like chromosomes entering cell division, are slightly condensed (about 20-fold) and so are laterally differentiated into alternate lightly and more heavily coiled regions. The effect of polytenisation is to amplify these differences in lateral structure so that each heavily coiled region forms a deeply staining band (chromomere), while the more lightly coiled regions form lightly staining interbands. Different bands, each representing a particular pattern of chromatin packing, differ greatly in size and appearance, and even a short segment of a polytene chromosome has a very specific and identifiable pattern of banding (Figure 3.8).

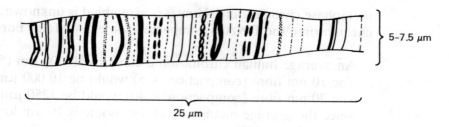

5–7.5 µm

25 µm

Figure 3.8 *Part of chromosome III from a salivary gland cell of* D. melanogaster

In *Drosophila melanogaster* over 5000 bands have been recorded, each containing, on the average, about 30 kb of DNA; it is probable that the average band includes between 5 and 10 genes.

(c) Polytenisation enables the construction of cytological maps showing the pattern of banding along each chromosome. Any changes in chromosome structure, such as deletion, duplication, inversion or translocation, can be recognised by alterations in the pattern of banding; in some instances particular gene mutations can be correlated with a specific band.

Further information

The heterochromatic regions flanking the centromeres are late replicating and are not amplified in these chromosomes. In *Drosophila* the heterochromatic regions of all four chromosome pairs associate together, forming a diffuse central region, the **chromocentre**.

The largest chromosome of *D. melanogaster* is chromosome III. At metaphase it is only about 2.5 µm long and contains just over 2 cm of fully extended DNA. The corresponding polytene chromosome is 943 µm long (excluding the centromeric region), is 5–7.5 µm wide and contains just over 2000 bands.

Question 3.5

Explain what is meant by dosage compensation. How is this achieved in mammals and *Drosophila*?

Answer 3.5

The two sets of chromosomes in a diploid organism constitute a finely balanced system, and if this balance is upset, by the presence of an extra autosome or by the absence of one member of a pair of homologous autosomes, the result is frequently a highly abnormal individual. Yet in the placental mammals the females have two X chromosomes, while the males are hemizygous and have only one X, which suggests that there is a mechanism which can compensate for this difference in gene dosage. In mammals this imbalance is restored by the inactivation of one of the X chromosomes present in females. This X becomes tightly coiled and is seen as a heterochromatic body (the **Barr body**) lying just below the nuclear membrane; in this state the X chromosome is wholly or partially inactive, so that both males and females have only one *active* X chromosome and the same genic balance. This phenomenon is known as **dosage compensation**.

It appears that the process of X inactivation occurs at random in the somatic cells during early development, so that in some cells it is the paternal X which is inactivated and in other cells the maternal X; further, once inactivation has occurred, all the progeny cells have the same inactive X.

In *Drosophila melanogaster* a different form of dosage compensation occurs. The X chromosomes are never condensed into Barr bodies, as in mammals, but certain genes on the male X chromosome are twice as active as the corresponding genes on each of the X chromosomes present in a female, so that the same amounts of the gene products are present in both sexes. Only certain genes are regulated in this way, and for other genes there is twice as much gene product present in females as in males.

NOTE

The Barr body is also known as the **sex chromatin** body.

Question 3.6

Describe evidence suggesting that X chromosome inactivation in female mammals occurs at random.

Answer 3.6

The implication of random X inactivation occurring during somatic development is that all female mammals are genetic mosaics, as in some patches of tissue the paternal X and in other patches the maternal X will be active.

(1) The tortoise-shell cat is circumstantial evidence for this hypothesis. Female cats heterozygous for *B* (black coat colour) and *b* (orange coat) have coats consisting of a mixture of black and orange blotches. The probable explanation is that the black and orange patches are patches of tissue where either the X chromosome carrying *b* or the one carrying *B* have been randomly inactivated.

(2) In humans a locus on the X chromosome encodes the enzyme glucose-6-phosphate-dehydrogenase (G6PD). The normal wild type allele encodes a fully active enzyme but a mutant allele encodes a variant form that is inactive but can be detected serologically. Fibroblasts can be isolated from a heterozygous female and grown in tissue culture, and the different clones can be tested for the presence of the active enzyme and the variant inactive form. About one-half of the clones grown from these fibroblasts produce the active enzyme, and the remainder produce the inactive variant; both proteins are *never* found in the same clone.

These results are only expected if there is only one active X in each clone and if there was random inactivation during development.

Question 3.7

What are 'cohesive ends' and what is their importance to phage lambda?

Answer 3.7

A linear molecule of DNA has **cohesive** ('sticky') ends when it has protruding 5′ single-stranded ends with complementary nucleotide sequences (Figure 3.9).

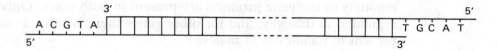

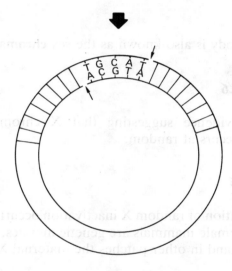

Figure 3.9 *Cohesive ends. These enable a linear molecule to circularise by complementary base pairing between the single-stranded extremities. The two single-strand gaps (arrowed) are sealed by DNA ligase*

A molecule with cohesive ends can circularise by complementary base pairing between the two single-stranded tails; this leaves two single-strand gaps in the sugar–phosphate backbone chains and these are sealed by DNA ligase, producing a covalently closed circular molecule (ccc DNA).

The importance to phage lambda of cohesive ends is that the DNA packaged into the phage heads and subsequently injected into a host cell is linear. This linear DNA must circularise before it can either replicate in or lysogenise the host cell (see Section 6.2.b), and this is possible because of the cohesive ends.

NOTES

1 The cohesive ends of lambda are 12 nucleotides long and have the sequences

_____ _ _ _____ CCCGCCGCTGGA 5'

5' GGGCGGCGACCT ———————— _ _ ————————

2 Linear molecules can be formed from circles by the reverse process.

3 When lambda replicates, the immediate product of replication is a con-catenate, 8 or more genomes long (see Figure 2.9). This concatenate is cut by a special phage-encoded enzyme which recognises the cohesive end sequences and makes a single-strand cut at each 5' end:

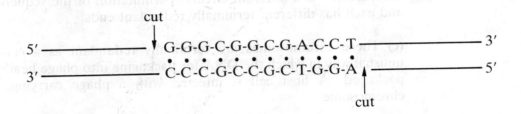

This simultaneously cuts the DNA into unit genome lengths and regenerates the cohesive ends.

Question 3.8

The chromosome of coliphage T2 is a linear molecule of DNA and it is both terminally redundant and circularly permuted. Explain the meaning of these terms and describe how these molecules are produced.

Answer 3.8

(a) Terminally redundant DNA has the sequence of nucleotides at one end of the molecule exactly repeated at the other end

5' CATTA ——————————————————— CATTA 3'

3' GTAAT ——————————————————— GTAAT 5'

If ABCDEFGH represents a unit set of genetic information, a terminally redundant sequence can be represented

ABCDEFGHAB

(*Note:* The chromosome of T2 is about 2×10^5 bp long and the terminally redundant sequences are between 1000 and 6000 bp each.)

43

(b) Different phage particles have different terminally redundant segments — the DNA from four different phages could be

ABCDEFGHAB
CDEFGHABCD
EFGHABCDEF
GHABCDEFGH

These chromosomes are different circular permutations of each other; in T2 all possible types of circular permutation occur and at equal frequencies. Each chromosome is a different circular permutation of the sequence ABCDEFGH and each has different terminally redundant ends.

(c) These chromosomes are terminally redundant as a consequence of the unusual way in which the DNA for packaging into phage heads is produced and packaged. A host cell is infected with a phage carrying, let us say, the chromosome

ABCDEFGHAB

First, replication produces many copies of this chromosome

ABCDEFGHAB, ABCDEFGHAB, ABCDEFGHAB, ABCDEFGHAB, etc.

Second, recombination (crossing-over) occurs between the terminally redundant ends of these copies of the chromosome, producing a long length of DNA many genomes long — a **concatenate:**

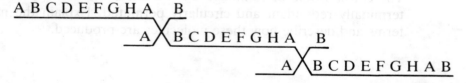

and so on, producing

A B C D E F G H A B C D E F G H A B C D E F G H A B C D E F G H . . .

Third, this DNA is packaged into phage heads: each phage head cuts off from the concatenate a length of DNA that exactly fills it: this piece of DNA is greater than unit length — 10 letters long in this example, compared with the 8 letters of a unit genome. Cutting off sequential 10 letter lengths produces molecules that are both terminally redundant and circularly permuted:

ABCDEFGHAB, CDEFGHABCD, EFGHABCDEF, etc.

NOTES

1 This method for packaging phage DNA is called headful packaging.

44

2 Crossing-over between the terminally redundant ends of the phage DNA is not formally different from crossing-over between two sister chromatids at meiosis, except that only one of the two reciprocally related products is recovered.

3.4 Supplementary Questions

3.1 Two unrelated bacterial species are grown synchronously in medium containing *low*-activity tritiated ($[^3H]$) thymidine. Half-way through the replication cycle, the cells were fed with *high*-activity tritiated thymidine. After the completion of the replication cycle, the DNA was extracted and autoradiographed. What can you deduce about the modes of replication from the following autoradiographs?

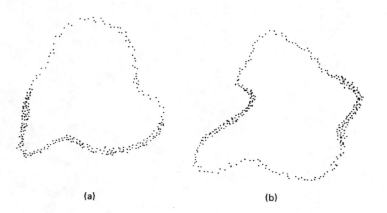

(a)   (b)

3.2 How would you modify the method used in the last experiment to show that the chromosome of *E. coli* replicates bidirectionally?

3.3 Naked chromosomes extracted from phage MS2 can infect *E. coli* protoplasts (bacteria with the cell wall removed) and result in the production of phage particles identical with the infecting phage. This infectivity is abolished if the chromosomes are first treated with RNase, but not with DNase: furthermore, if the infecting RNA is labelled, the label never appears among the progeny phage. Account for these observations.

3.4 DNA extracted from phage lambda is treated with two exonucleases which only attack single-stranded DNA. Exonuclease A digests DNA only from exposed 5' ends, while exonuclease B only attacks free 3' ends. What effect would this treatment have on the biological activity of the lambda DNA?

3.5 Distinguish between (a) chromatin, (b) chromosome, (c) chromatid and (d) chromomere.

3.6 A few of the *Vicia* chromosomes observed by Taylor (see Question 3.2) were labelled as shown overleaf. How and when did these chromosomes arise?

3.7 How does constitutive heterochromatin differ from euchromatin?

3.8 When chromatin is extracted from eucaryotic cells, treated with endonuclease and the DNA is separated from the chromatin, the resulting fragments of DNA are not, as might be expected, random but are either 200 bp or a multiple of 200 bp long. What is the significance of this observation?

4 The transmission of genes I: Mendelian ratios in eucaryotes

4.1 Definitions and Concepts

In Chapters 1 and 2 we saw that each eucaryotic chromosome consists of a single very long molecule of DNA, together with associated proteins, and that the genetic information is stored in the sequence of nucleotides along this DNA. Furthermore, each chromosome, and hence each molecule of DNA, is differentiated along its length into a large number (perhaps between 500 and 2000) of units of genetic function, the **genes**. It is how these genes are inherited that concerns us here. Although we shall place considerable emphasis on genetic ratios (the proportions of the different types of progeny arising from a cross), it is important not to forget the underlying principles of gene segregation, assortment and interaction which are responsible for these ratios.

Since the chromosomes occur in pairs, the genes they carry must also occur in pairs. However, the two genes at a particular locus on each of a pair of homologous chromosomes need not be identical, as **mutation**, a process which alters the nucleotide sequence of a gene (Chapter 11), can change the way in which a gene is expressed. These different forms of the same gene are **alleles**, and, in many instances, the altered or **mutant allele** can no longer produce the active gene product; in contrast, the allele normally present and producing a fully active gene product is the **wild type** allele. The terms 'gene', 'allele' and 'marker' tend to be used interchangeably.

Gene symbols are commonly an abbreviation or an acronym of the mutant character — thus, in *Drosophila* the gene responsible for the character 'vestigial wing' is symbolised vg (or very infrequently vg^-) and the corresponding wild-type allele is variously indicated as $+$, vg^+ or Vg. The conventions used differ widely, particularly from one organism to another, but, in general, a capital letter or a $(+)$ or a $(+)$ superscript indicates a wild type allele and a lower case letter(s) or a $(-)$ superscript a mutant allele.

When both members of a pair of homologous chromosomes carry the same allele (AA or aa), the cell is **homozygous** and the organism is a **homozygote**; when the two alleles are different (Aa), the cell is **heterozygous** and the organism is a **heterozygote**.

The genetic description or constitution of an organism is its **genotype** and its observable properties or appearance is its **phenotype**.

If the *AA* and *Aa* genotypes have the same wild phenotype, then *A* is said to be **dominant** over *a*, or *a* **recessive** to *A*; usually, but not always, the wild type allele is dominant and the mutant allele is recessive. This is most simply seen if we consider a *Aa* heterozygote where the *A* allele results in the production of an active gene product, whereas the *a* mutant allele fails to produce this product; so long as the *A* gene is present, the active gene product will be produced and both *AA* and *Aa* individuals will have a wild phenotype and only *aa* individuals will have a mutant phenotype.

It is important to remember that because an organism has a particular wild type gene does not mean that the corresponding phenotype will necessarily be displayed. Although most seedlings have all the genes necessary to produce chlorophyll and to turn green, a seedling grown in the dark will remain pale yellow and will only turn green if it is transferred to light; this is because light is also necessary for the production of the green pigment. Indeed, for nearly all inherited characters the phenotype is the result of interactions between the genotype and the environment.

4.2 The Principle of Segregation

In 1866 the Catholic priest **Gregor Mendel** published the results of his experiments on plant hybridisation and correctly stated the principles of inheritance — a quite remarkable achievement, since, at that time, nothing was known about heredity and the chromosomal basis of inheritance had not been discovered. The importance of his work is that he recognised that it was *not* the characters themselves that were inherited but the determinants of characters — what we now call genes.

The segregation of genes is best understood if you remember the behaviour of the chromosomes at meiosis — in particular, that they occur in homologous pairs and that a copy of just one homologue passes into every gamete; furthermore, two of the four meiotic products receive a copy of one homologue and the remaining two a copy of the other homologue. Because meiosis is such a precise and regular process, a *Aa* homozygote will produce one-half *A* and one-half *a* gametes (Figure 4.1).

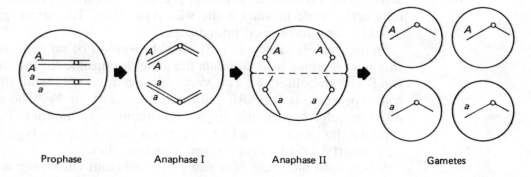

Prophase   Anaphase I   Anaphase II   Gametes

Figure 4.1 *Independent segregation. In a* Aa *heterozygote half the gametes produced at meiosis are* A *and half are* a

Thus, at meiosis the two members of a pair of alleles (*A* and *a*) will segregate into different gametes; they are independently transmitted. This is the **Principle of Segregation** (sometimes referred to as Mendel's first law), first stated by Mendel in 1866 and illustrated by the following crosses.

One stock of *Drosophila melanogaster* has vestigial wings and cannot fly: it is homozygous for a recessive gene (*vg*) on chromosome II. Wild type stocks are homozygous for the wild type (*vg*$^+$ or +) allele. If we cross these true breeding parental (P) stocks, all the next generation, known as the **first filial** or **F1** generation, will be heterozygotes and will have normal wings:

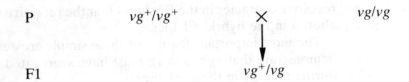

P vg^+/vg^+ × vg/vg

F1 vg^+/vg

(1) If these F1 flies are now **backcrossed** to the *vg/vg* parent, then the heterozygous parent will produce half *vg*$^+$ and half *vg* gametes, while the vestigial parent will only produce *vg* gametes. Thus, on the average, half the progeny are *vg*$^+$/*vg* and have normal wings and half are *vg/vg* and have vestigial wings: there is a 1:1 ratio among the progeny (Figure 4.2a). This is the **second filial** or **F2** generation.

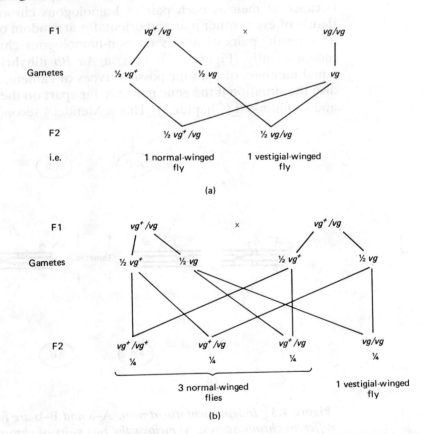

Figure 4.2 *The segregation of a pair of alleles in* D. melanogaster. *F1* vg$^+$/vg *flies are either backcrossed to the* vg/vg *parent (i), or crossed among themselves (ii)*

(2) If instead the F1 flies were crossed among themselves,

$$vg^+/vg \times vg^+/vg,$$

then each parent would produce equal numbers of vg^+ and vg gametes and, among the F2 progeny, there would be a ratio of three wild type to one vestigial winged (Figure 4.2b).

Note that in the F2 there are three different genotypes but only two different phenotypes; the vg^+/vg^+ and vg^+/vg genotypes can only be distinguished by making further crosses (Question 4.2). Furthermore, the reappearance of the recessive character in the F2 shows that the recessive vg gene was neither lost nor altered in the hybrid F1 flies.

The most important feature of these simple crosses is that the observed ratios demonstrate that vg^+ and vg must have segregated randomly into the gametes during meiosis in these F1 flies.

4.3 Independent Assortment

If two pairs of alleles located on different pairs of chromosomes are segregating, then each pair of alleles will segregate independently of the other pair. This is because at meiosis each pair of homologous chromosomes behaves independently of every other pair and orientates at random on the metaphase plate and, as a result, pairs of alleles on non-homologous chromosomes will also assort independently (Figure 4.3); thus, a $Aa\ Bb$ **dihybrid** will produce statistically equal numbers of the four possible types of gamete, AB, $Ab\ aB$ and ab. This is also the situation if the gene pairs are far apart on the same pair of chromosomes and so unlinked (Chapter 5). This is Mendel's second principle, the **Principle of**

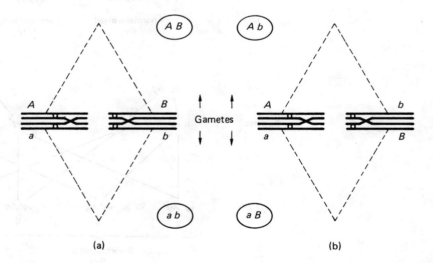

(a)   (b)

Figure 4.3 *Independent assortment. A–a and B–b are two pairs of genes located on different chromosomes. At meiosis the two pairs of chromosomes (bivalents) orientate at random on the metaphase plate so that the two arrangements (a) and (b) occur at equal frequency and equal numbers of the four types of gamete will be produced*

Independent Assortment, and it is illustrated by the 1:1:1:1 and 9:3:3:1 ratios; these are really combinations of two separate 1:1 and 3:1 ratios, respectively.

Ebony flies of *Drosophila melanogaster* have darker than normal bodies and are homozygous for a recessive gene *e* on chromosome III. If ebony flies are crossed with a vestigial-winged stock, then all the F1 flies will be $vg^+/vg\ e^+/e$ dihybrids (the two separate fraction bars indicate that these loci are on different chromosomes) and these will produce four types of gamete, $vg^+\ e^+$, $vg^+\ e$, $vg\ e^+$ and $vg\ e$, in equal numbers.

(1) If these flies are backcrossed to a stock with ebony bodies and vestigial wings, i.e.

$$vg^+/vg\ e^+/e\ \times\ vg/vg\ e/e$$

then the four progeny genotypes and phenotypes will be

$vg^+/vg\ e^+/e$	$vg^+/vg\ e/e$	$vg/vg\ e^+/e$	$vg/vg\ e/e$
normal wings, normal bodies	normal wings, ebony bodies	vestigial wings, normal bodies	vestigial wings, ebony bodies

in a   1   :   1   :   1   :   1   ratio

Note that because the $vg/vg\ e/e$ parent only produces $vg\ e$ gametes, the progeny phenotypes are the same as the genotypes of the gametes produced by the dihybrid parent.

(2) If the F1 flies are crossed among themselves (a **dihybrid cross**), i.e.

$$vg^+/vg\ e^+/e\ \ \ \ \times\ \ \ \ vg^+/vg\ e^+/e$$

then each parent will produce four types of gamete and the male and female gametes will unite at random in all possible combinations; these 16 (4 × 4) gametic combinations produce nine different genotypes distributed among four progeny phenotypes; these four phenotypes occur in a 9:3:3:1 ratio (Figure 4.4), so confirming that the pairs of alleles are segregating independently.

Note that this ratio, or any other ratio, can be more simply derived by the use of a **branch diagram**. Each pair of alternative phenotypes occurs in a 3:1 ratio in the F2 and, if the gene pairs are segregating independently, there will be a 3:1 ratio of normal:ebony body among both the normal-winged and vestigial-winged flies. Thus, among the progeny,

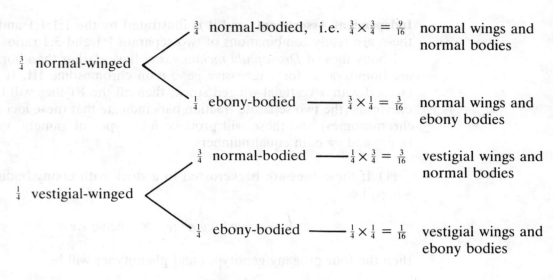

$\frac{3}{4}$ normal-bodied, i.e. $\frac{3}{4} \times \frac{3}{4} = \frac{9}{16}$ normal wings and normal bodies

$\frac{3}{4}$ normal-winged

$\frac{1}{4}$ ebony-bodied $\quad\frac{3}{4} \times \frac{1}{4} = \frac{3}{16}$ normal wings and ebony bodies

$\frac{3}{4}$ normal-bodied $\quad\frac{1}{4} \times \frac{3}{4} = \frac{3}{16}$ vestigial wings and normal bodies

$\frac{1}{4}$ vestigial-winged

$\frac{1}{4}$ ebony-bodied $\quad\frac{1}{4} \times \frac{1}{4} = \frac{1}{16}$ vestigial wings and ebony bodies

$vg^+/vg \quad e^+/e \quad \times \quad vg^+/vg \quad e^+/e$

Female Gametes	Male Gametes			
	$vg^+\,e^+$	$vg^+\,e$	$vg\,e^+$	$vg\,e$
$vg^+\,e^+$	$vg^+/vg^+\ e^+/e^+$ Wild type	$vg^+/vg^+\ e^+/e$ Wild type	$vg^+/vg\ e^+/e^+$ Wild type	$vg^+/vg\ e^+/e$ Wild type
$vg^+\,e$	$vg^+/vg^+\ e^+/e$ Wild type	$vg^+/vg^+\ e/e$ Ebony	$vg^+/vg\ e^+/e$ Wild type	$vg^+/vg\ e/e$ Ebony
$vg\,e^+$	$vg^+/vg\ e^+/e^+$ Wild type	$vg^+/vg\ e^+/e$ Wild type	$vg/vg\ e^+/e^+$ Vestigial	$vg/vg\ e^+/e$ Vestigial
$vg\,e$	$vg^+/vg\ e^+/e$ Wild type	$vg^+/vg\ e/e$ Ebony	$vg/vg\ e^+/e$ Vestigial	$vg/vg\ e/e$ Vestigial, ebony

Thus, for every 16 flies:

9 will be wild type (4 different genotypes)
3 will have vestigial wings and normal bodies (2 genotypes)
3 will have normal wings and ebony bodies (2 genotypes)
1 will have vestigial wings and ebony body

Figure 4.4 *A dihybrid cross in* D. melanogaster. *This type of diagram, setting out all the possible combinations of the male and female gametes, is known as a Punnett square*

4.4 Gene Interaction

In a generalised dihybrid cross (Section 4.3) or selfing (*A/a B/b* × *A/a B/b* or *A/a B/b* ⊕ where ⊕ is the symbol for self cross), there is, provided that the two pairs of alleles affect widely contrasting characters, a 9:3:3:1 ratio of progeny phenotypes, viz.

9 AB	(individuals heterozygous or homozygous for both *A* and *B*)
3 Ab	(individuals heterozygous or homozygous for *A* but homozygous for *b*)
3 aB	(individuals heterozygous or homozygous for *B* but homozygous for *a*)
1 ab	(individuals homozygous for both *a* and *b*)

but when both loci affect the expression of the SAME character, there may be a consistent modification of the usual 9:3:3:1 ratio. When the different loci **interact** in the production of a particular phenotypic character, the 9:3:3:1 can be modified to 9:7, 9:3:4, 13:3, 9:6:1, 12:3:1 or 15:1. In backcrosses the expected 1:1:1:1 ratio is correspondingly modified.

What is important is not the ratio itself but the particular way in which the gene products have interacted to produce that ratio; conversely, if we observe a certain ratio, it is possible to make some deductions about how the loci interact.

We shall illustrate gene interaction, using three examples in maize, *Zea mays*, all involving the production of pigment. In maize the four dominant alleles A_1, A_2, *C* and *R* MUST be present before any anthocyanin pigment can form in the aleurone of the grain and in certain other parts of the plant, and when all four of these genes are present, the aleurone is normally red; if another dominant gene, *Pr*, is present, this red pigment is converted to a purple pigment. These five loci assort independently of each other. A_1, A_2, *C*, *R* and *Pr* control sequential steps in the pathway leading to anthocyanin production (biochemical pathways are explained in Chapter 7):

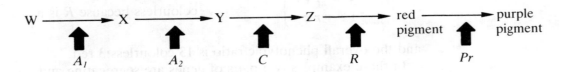

where W, X, Y and Z are colourless intermediates and the product of the A_1 gene converts intermediate W into intermediate X, that of the *R* gene converts intermediate Z into a red pigment, and so on.

(a)   Complementary Genes — the 9:7 Ratio

Consider the selfing *Cc Rr* ⊕ (where A_1 and A_2 are homozygous and not segregating). The progeny phenotypes will be

9	C R	red aleurone
3	C r	
3	c R	colourless because either *C* or *R* is absent
1	c r	

This produces a 9:7 ratio of red:colourless and occurs because BOTH dominants are required for red pigment to form. *C* and *R* are called **complementary genes**.

(b) Epistasis — the 9:3:4 Ratio

If the selfing is $Rr\ Pr\ pr\ \oplus$, the progeny phenotypes are

9	$R\ Pr$	purple aleurone
3	$R\ pr$	red aleurone
3	$r\ Pr$	colourless
1	$r\ pr$	

and there is a ratio of 9 purple : 3 red : 4 colourless. Red colour only forms if R is present, but if Pr is also present, then the red pigment is converted to purple. Both $r\ Pr$ and $r\ pr$ will be colourless, since homozygosity for the recessive r allele has prevented the expression of Pr — this is known as **recessive epistasis**.

(c) Inhibiting Genes — the 13:3 Ratio

At the C locus there is a third allele, C^I, and the three alternative alleles C, c and C^I are known as **multiple alleles**. Whenever C^I is present, it inhibits the activity of the C gene product. Thus, $C^I C^I$, $C^I C$, $C^I c$ and cc genotypes are all colourless and C^I is dominant over C.

In the selfing $Rr\ C^I C\ \oplus$ the progeny phenotypes will be

9	$R\ C^I$	colourless, C^I present
3	$R\ C$	red aleurone — both C and R present
3	$r\ C^I$	colourless because R is absent
1	$r\ c$	

and the overall phenotypic ratio is 13 colourless:3 red.

In these examples two pairs of genes are segregating and, as a result of gene interaction, the standard dihybrid ratio (9:3:3:1) is modified. Thus, if we observe one of these modified ratios in a cross where the pattern of inheritance is not known, it is a good indication that two pairs of genes are segregating.

Note that the 9:7 and 13:3 ratios may not be immediately recognisable as such, as they closely approximate to the 1:1 and 3:1 ratios. However, a 9:7 and 1:1 are unlikely to be confused, as selfing a Aa heterozygote produces a 3:1 and not a 1:1 ratio. The 13:3 and 3:1 are more likely to be confused, but there are seven different genotypes among the colourless progeny in the former and only one among the latter, a situation easily detected by backcrossing the colourless progeny to the parent stocks. Furthermore, there are simple statistical tests which enable us to determine whether an observed ratio is in good or poor agreement with each of the possible theoretical ratios.

4.5 Sex Linkage

In Chapter 1 we saw that sex is frequently determined by a special pair of sex chromosomes and that in both *Drosophila* and placental mammals the females have a pair of X chromosomes and the males one X and a dissimilar Y chromosome. Genes that are carried on the X chromosome are said to be sex-linked, as they are not inherited independently of sex. This is because the females inherit an X from each parent, while the males always inherit an X from their mothers and a Y from their fathers. This results in a criss-cross pattern of inheritance (see Question 4.10).

A well-known example of a sex-linked recessive condition is haemophilia A (bleeders' disease). The harmful recessive allele (*h*) is quite rare and is usually transmitted through heterozygous females (carriers of the recessive allele) who marry normal males and pass the harmful gene on to their sons; on the average, half of the sons of a carrier female will be **hemizygous** for the *h* allele and will be haemophiliacs. Many of the affected males die comparatively young, but if they marry and have children, then all their daughters will be carriers and all their sons will be normal (Figure 4.5).

Because of the severity of the condition and the rarity of the *h* allele, affected females are extremely rare, as they would have to have a carrier mother AND an affected father. The result is that the character appears to be transmitted through the females of one generation to the males of the next generation.

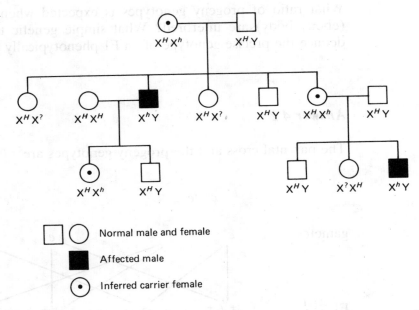

Figure 4.5 *A human pedigree for haemophilia A. The genotypes have been deduced from the pedigree. X? indicates that the allele present on that particular X chromosome cannot be deduced from the information given*

4.6   Questions and Answers

Question 4.1

Explain what you understand by
   (a) segregation;
   (b) independent assortment.

Answer 4.1

(a) Segregation refers to the way that the two members of a pair of homologous chromosomes, or the two members of a pair of alleles, pass into different gametes at meiosis.

(b) If two pairs of alleles A, a and B, b are on non-homologous chromosomes, then A, a and B, b will segregate independently of each other. This is because, at meiosis, every pair of homologous chromosomes aligns at random on the metaphase plate and so behaves independently of every other pair — thus, A, a and B, b will assort independently and there will be equal numbers of AB, Ab aB and ab gametes.

Question 4.2

What ratio of progeny genotypes is expected when flies heterozygous for e (ebony body) are interbred? What simple genetic test would enable you to deduce the precise genotype of an F1 phenotypically wild-type fly?

Answer 4.2

The parental cross and the progeny genotypes are

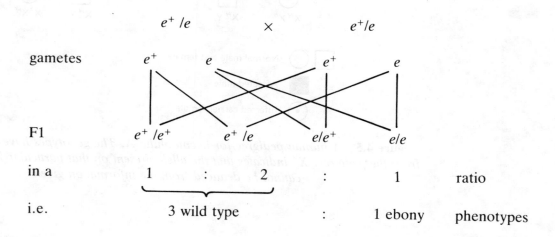

		e^+/e	$\times$	e^+/e		
gametes	e^+	e		e^+	e	
F1	e^+/e^+	e^+/e		e/e^+	e/e	
in a	1	:	2	:	1	ratio
i.e.		3 wild type		:	1 ebony	phenotypes

Thus, among the phenotypically wild type progeny one-third are e^+/e^+ and two-thirds are e^+/e.

The simplest way to determine whether a wild type fly is e^+/e^+ or e^+/e is to make a **test** cross and backcross the fly to an ebony-bodied fly.

The cross can be either $e^+/e^+ \times e/e$ or $e^+/e \times e/e$.

If the fly under test was e^+/e^+, then all the progeny will be wild type, but if it was e^+/e, then one-half of the progeny will be wild type and one-half ebony.

NOTE

In a test cross an individual (or stock) with an unknown or unconfirmed genotype is crossed with an individual (or stock) homozygous recessive at all the loci of interest. Thus, all the genes contributed by the F1 individual under test will be expressed among the test-cross progeny.

Question 4.3

What ratio of progeny phenotypes is expected in the cross $A/a\ B/b \times A/a\ b/b$?

Answer 4.3

Consider only the segregation of A and a; the cross is $A/a \times A/a$ and there will be a 3:1 ratio of $A{:}a$ progeny phenotypes.

Consider only the segregation of B and b; the cross is $B/b \times b/b$ and there will be a ratio of 1:1 $B{:}b$ progeny phenotypes.

There is a 1:1 ratio superimposed on a 3:1 ratio. Thus, among the three-quarters of the progeny with an A phenotype, one-half will be B and one-half will be b and, overall, three-eighths will be AB (i.e. $\frac{3}{4} \times \frac{1}{2}$) and three-eighths will be Ab. Among the one-quarter of the progeny with an a phenotype, one-half will be B and one-half will be b and, overall, one-eighth ($\frac{1}{4} \times \frac{1}{2}$) will be aB and one-eighth will be ab. There will be a 3:3:1:1 ratio among the progeny phenotypes.

An alternative method is to construct a Punnett Square (Figure 4.6).

Gametes from A/a b/b parent	Gametes from A/a B/b parent			
	AB	aB	Ab	ab
Ab	A/A B/b (AB)	A/a B/b (AB)	A/A b/b (Ab)	A/a b/b (Ab)
ab	A/a B/b (AB)	a/a B/b (aB)	A/a b/b (Ab)	a/a b/b (ab)

Figure 4.6 *Progeny from the cross* A/a B/b × A/a b/b. *The F1 phenotypes are shown in parentheses*

Summing the progeny phenotypes: 3 AB
3 Ab
1 aB
1 ab

Question 4.4

In maize A_1 C and R are all required for the production of anthocyanin in the aleurone. What proportion of coloured progeny are expected in the following crosses?

(a) $A_1 \, a_1 \, Cc \, Rr \oplus$

(b) $A_1 \, a_1 \, Cc \, Rr \times A_1 \, a_1 \, CC \, Rr$

(c) $A_1 \, a_1 \, Cc \, Rr \times a_1 \, a_1 \, cc \, Rr$

Answer 4.4

(a) All three gene pairs are segregating in a 3:1 ratio, so the proportion of coloured progeny (A_1 C R)

$$= 3/4 \times 3/4 \times 3/4 = 27/64$$

(i.e. a 3:1 superimposed on a 9:7).

(b) A_1/a_1 and R/r segregate in a 9:7 ratio and $\frac{9}{16}$ of the progeny will be A_1 R. Since all the progeny inherit C from the male (right-hand) parent, the proportion of coloured progeny is $\frac{9}{16}$.

(c) One-half of the progeny receive A_1; one-half of the progeny receive C; three-quarters of the progeny receive R; and the proportion of coloured progeny is $\frac{1}{2} \times \frac{1}{2} \times \frac{3}{4} = \frac{3}{16}$.

NOTE

In the formula for a cross it is usual to put the genotype of the female parent at the left.

Question 4.5

The fruits of the summer squash, *Cucurbita pepo*, are spherical, discoid or elongated. When two different *spherical* races were crossed, the F1 all had *discoid* fruits; when these F1 plants were selfed, the F2 segregated 53 *discoid*, 37 *spherical* and 4 *elongate*. How can you explain these results?

Answer 4.5

The appearance of a third phenotype in the F2 shows that at least two genes are segregating.

It is probable that 53:37:6 is a modified 9:3:3:1 ratio; the observed ratio is very close to 9:6:1 (expect 51.2:34.1:5.8).

In order to obtain a modified 9:3:3:1 in the F2, the F1 plants must have been the double heterozygotes (i.e. *Aa Bb*) and the pure breeding *spherical* parents were *AA bb* and *aa BB*.

Thus, the F2 progeny are 9 *A B* with *discoid* phenotypes,

3 *A b*
3 *a B* } with *spherical* phenotypes,

1 *a b* with an *elongate* phenotype.

Thus, in the presence of *either A or B* the phenotype is *spherical*, but if *both* are present, the fruit becomes flattened and *discoid*.

NOTE

The parental strains could not have been *AA BB* and *aa bb*, as then one would have been discoid and the other spherical.

Question 4.6

Explain, with examples, what is meant by

(a) dominance,

(b) partial (incomplete) dominance,

(c) Co-dominance.

Answer 4.6

(a) **Dominance** refers to the phenotype expressed in a heterozygote. If *AA* and *Aa* genotypes both have a *A* phenotype, then *A* is the dominant phenotype (and allele) and *a* the recessive phenotype (and allele). In *Drosophila* both vg^+/vg^+ and vg^+/vg flies have normal wings — vg^+ is dominant.

(b) **Partial dominance** is where the heterozygote has an intermediate phenotype. For example, in carnations (and many other plants) *RR* plants have red flowers, *rr* plants have white flowers and the *Rr* heterozygotes are pink-flowered. This is because the *RR* genotypes produce twice as much red pigment as *Rr* plants; *rr* genotypes produce no red pigment at all.

(c) **Co-dominance** is where *both* alleles of a gene pair are expressed and produce a detectable gene product. In normal humans the red blood cells contain a protein, haemoglobin A or HbA, which is essential for transporting oxygen around the body; these individuals have the genotype *HbᴬHbᴬ*. Individuals suffering from sickle cell anaemia only have abnormal HbS and the RBCs become sickle-shaped and are unable to transport oxygen; these individuals are *HbˢHbˢ*, suffer from a progressive haemolytic anaemia and die at an early age.

The $Hb^A Hb^S$ heterozygotes have *both* types of haemoglobin and, although they are carriers of the harmful gene, they themselves are largely unaffected.

Question 4.7

Write a short account of multiple alleles.

Answer 4.7

Although any diploid individual only carries two alleles of a particular gene (one on each homologous chromosome), there may be many different alleles in the population as a whole. These are referred to as multiple alleles, and the group of alternative alleles forms a **multiple allelic series**.

A simple example, with only three common alleles involved, is the ABO blood group system in humans. There are four blood groups in this system, determined by three alleles at a single locus I (I stands for isoagglutinogen, another name for an antigen) (Table 4.1).

Table 4.1

Blood group	Possible genotypes	Antigens present
O	$I^O I^O$	none
A	$I^A I^A$ and $I^A I^O$	A
B	$I^B I^B$ and $I^B I^O$	B
AB	$I^A I^B$	A and B

I^A and I^B encode the A and B antigens found on the red blood cells and in $I^A I^B$ individuals both antigens are present. I^O (or i) is the null allele and does not encode any antigen. I^A and I^B are co-dominant and both are dominant to I^O.

Coat colour in mice, rabbits and guinea-pigs is determined by a series of four alleles: C (full colour or grey), c^{ch} (chinchilla, less intense pigmentation), c^h (Himalayan, albino coat but with ears, tail, feet and nose pigmented) and c (full albino). C is dominant to each of the other three alleles, c^{ch} is dominant to both c^h and c, and c^h is dominant to c.

Another well-known example where phenotype is controlled by a series of multiple alleles is at the white-eye (w) locus in *Drosophila*. Wild type (w^+) flies have bright red eyes and ww flies have colourless (white eyes); between these extremes there are many different shades and intensities of yellow, brown, orange and red pigmentation, determined by a series of multiple alleles at the w locus — for example, w^a, apricot; w^{bf}, buff; w^e eosin; and w^c, cherry.

Question 4.8

(a) List the possible marriages whereby a group A father can have a group O child. In each marriage what proportion of the children are expected to be group O?

(b) Could a group B child have a group A father? Explain.

Answer 4.8

(a) Since the child is group O, it has the genotype $I^O I^O$. Thus, the father must contribute an I^O allele, and, since he is group A, must be I^A, I^O. The mother also must contribute I^O and her genotype could be $I^A I^O$, $I^B I^O$ or $I^O I^O$.

The possible marriages are:

$I^A I^O \times I^A I^O$ ($\frac{1}{4}$ of the children will be $I^O I^O$)
$I^B I^O \times I^A I^O$ ($\frac{1}{4}$)
$I^O I^O \times I^A I^O$ ($\frac{1}{2}$)

(b) Yes, provided that the group A father was $I^A I^O$ and contributed I^O to the child. The mother must contribute I^B, but could be $I^B I^B$, $I^B I^O$ or $I^A I^B$.

Question 4.9

If $c^h c$ Himalayan and $C c^{ch}$ wild type rabbits are crossed, what proportions of progeny phenotypes are expected (use the information from Question 4.6)?

Answer 4.9

There are 4 progeny genotypes and these occur in a 1:1:1:1 ratio:

$C c^h$ 1 $\Big\}$ 2 wild type progeny, since C is dominant over both c^h and c
$C c$ 1

$c^{ch} c^h$ 1 $\Big\}$ 2 chinchilla, since c^{ch} is dominant over c
$c^{ch} c$ 1

i.e. there will be 1 wild type to 1 chinchilla among the progeny.

Question 4.10

In *Drosophila* the white-eye locus (w) is on the X chromosome and $X^w X^w$ females and $X^w Y$ males have white eyes. If white-eyed females are mated to wild type (red-eyed) males, what F1 progeny are expected? What proportions of progeny phenotypes are found in the F2 if these F1 flies are allowed to breed among themselves?

Answer 4.10

The question is best answered in the form of a diagram:

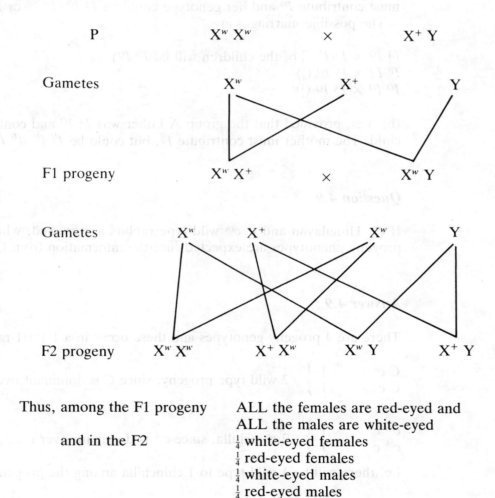

| P | $X^w X^w$ | × | $X^+ Y$ |

Gametes X^w X^+ Y

F1 progeny $X^w X^+$ × $X^w Y$

Gametes X^w X^+ X^w Y

F2 progeny $X^w X^w$ $X^+ X^w$ $X^w Y$ $X^+ Y$

Thus, among the F1 progeny   ALL the females are red-eyed and ALL the males are white-eyed

and in the F2 $\frac{1}{4}$ white-eyed females
$\frac{1}{4}$ red-eyed females
$\frac{1}{4}$ white-eyed males
$\frac{1}{4}$ red-eyed males

NOTE

The F1 progeny phenotypes clearly show that white-eye is not transmitted independently of sex. Observe that the parental phenotypes reappear in the F2.

Question 4.11

In cats females homozygous for the dominant *B* allele are black and *bb* homozygotes are orange. When black and orange cats are mated, the female progeny are always 'tortoise-shell' and their coats show a mottling of small black and orange patches, while the male progeny have the same coat colours as their mothers. Only very rarely are male tortoise-shell cats found.

How do you explain these results? What progeny are expected if tortoise-shell females are mated with black males?

Answer 4.11

Since the F1 males have the same coat colour as their mothers, the *B* locus is clearly X linked. Hence, the parental cross and the F1 is either

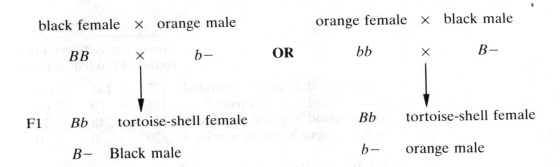

The tortoise-shell females are the consequence of the random inactivation of one of the X chromosomes. In some cell lines it is the X carrying *B* that is inactivated, resulting in a patch of orange tissue; in other cell lines it is the X carrying the recessive *b* allele that is inactivated, so that the *B* allele remains active and produces a patch of black tissue.

The very occasional tortoise-shell males are XXY individuals produced as a consequence of non-disjunction (Question 1.3); these, like females, have two X chromosomes, one of which is randomly inactivated.

The mating tortoise-shell female × black male is *Bb* × *B*− and produces equal numbers of the following progeny:

BB   black females
B−   black males
Bb   tortoise-shell females
b−   orange males

4.7 Supplementary Questions

4.1 Explain what is meant by dominance and recessiveness.

4.2 What is a reciprocal cross?

4.3 You have found a plant which, unusually, has white instead of the normal purple flowers? What experiments would you carry out to demonstrate simple Mendelian inheritance?

4.4 Deaf mutes are homozygous for a deleterious recessive gene. Occasional marriages between two deaf mutes produce only normal children. How do you explain this?

4.5 In humans albinism is due to homozygosity for a recessive gene. A woman whose father was an albino marries a man whose grandfather was similarly affected. What is the probability that (a) their first child will be an albino and (b) *both* members of a pair of non-identical twins would be albinos?

4.6 In peas (*Pisum sativum*) two pairs of alleles determine the seed characters green *v.* yellow and round *v.* wrinkled. In four separate crosses the following results were obtained:

Parents	Progeny			
	green round	green wrinkled	yellow round	yellow wrinkled
green round × yellow wrinkled	137	141	139	145	(i)
green round × green round	175	0	60	0	(ii)
green round × green wrinkled	221	248	73	87	(iii)
yellow round × green wrinkled	370	0	0	0	(iv)

(a) Which alleles are dominant?
(b) Write down the genotypes of the parents in each cross.

4.7 In a maternity hospital three babies are accidentally mixed up. The blood groups of the babies are (i) O, (ii) B and (iii) AB, and those of the three sets of parents (a) A and O, (b) A and AB and (c) AB and O. Which babies belong to which parents?

5 The transmission of genes II: Linkage

5.1 Linkage and Recombination

In Chapter 4 we saw that a *A/a B/b* double heterozygote, produced by crossing *A/A b/b* and *a/a B/B* mutants, produces four different types of gamete, *AB, Ab, aB* and *ab*, and that when the *A* and *B* loci are on different chromosomes, these four types of gamete are produced in equal numbers — the *A* and *B* loci segregate independently and are said to be **unlinked**. However, departures from independent assortment can occur when two (or more) loci which are segregating are on the same pair of homologous chromosomes. This is called **linkage**, and such linked genes do not segregate in the 1:1:1:1 and 9:3:3:1 ratios described.

Let us suppose that the *A* and *B* loci are linked and that the double heterozygote has been produced by crossing a wild type with a doubly homozygous recessive mutant:

$$P \quad \frac{A \ B}{A \ B} \times \frac{a \ b}{a \ b}$$

$$F1 \quad \frac{A \ B}{a \ b}$$

In writing the formula for this cross, the unbroken fraction bars indicate that the *A* and *B* loci are on the same chromosome.

At meiosis it is the pairs of chromosomes which segregate, not the individual genes, and so it might be expected that the two particular genes on the same chromosome would always segregate together, so that in this cross only *AB* and *ab* gametes would be produced and the *A* and *B* loci would show **complete linkage**. However, although there is a tendency for genes close together on the same chromosome to remain together when they enter the gametes, complete linkage is only rarely observed, because of the occurrence of **crossing-over** or **recombination**. The consequence of linkage is that there is an excess of the two types of gamete carrying the same combination of alleles as were present in the parental cross; in this example these **parental types** are the *AB* and *ab* gametes.

Crossing-over occurs at the prophase of meiosis after the homologous chromosomes have paired along their length and replicated (Figure 1.3), and it results in the physical exchange of genetic material between two non-sister chromatids (Figure 5.1).

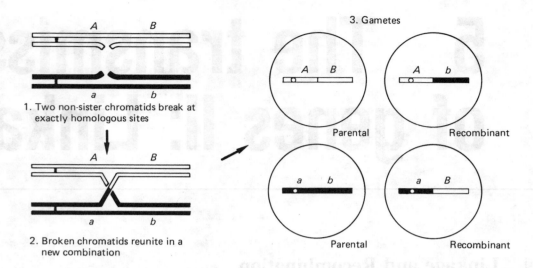

Figure 5.1 *Recombination. A single crossover between two loci on the same chromosome produces two parental type (AB and ab) and two non-parental type or recombinant (aB and Ab) gametes*

At the molecular level crossing-over is a complex process but the result is as though two non-sister chromatids have broken at exactly homologous points and then reunited in the alternative combination: note that each event involves only two of the four chromatids, although multiple crossing-over events can involve two, three or all four chromatids (Figure 5.2). At least one crossover must normally occur along the length of each pair of homologous chromosomes (i.e. along each bivalent), as otherwise the pairs of homologues will not be held together as a bivalent and will fail to orientate correctly on the metaphase plate: this results in the abnormal segregation of the chromosomes at anaphase.

It is important to realise that the chiasmata seen at meiotic prophase and metaphase do not show the actual process of crossing-over: this has already taken place but they are the relics, or visible manifestations, of this process.

5.2 Mapping by Recombination

Crossing-over occurs at random along the length of a bivalent and the frequency of crossing-over between two loci on the same chromosome is proportional to the genetic distance between them — the closer the two loci the less is the probability of a crossover occurring between them. Thus, the frequency of crossing-over is a measure of the genetic distance between two loci; because we can measure genetic distance in this way, it is possible to construct **chromosome** or **linkage maps** showing the orders of loci along chromosomes.

Consider the cross

$$\frac{A \; B}{a \; b} \times \frac{a \; b}{a \; b}$$

(more conveniently written *A B/a b × a b/a b*)

where the A and B loci are on the same chromosome (this is indicated by the single unbroken fraction bar in the formula). Note that it is always expedient to use the doubly recessive homozygote as one parent, as then crossing-over only has to be considered in the other parent and the progeny phenotypes directly indicate the genotypes of the gametes formed by the heterozygous parent.

The A B/a b parent will produce four types of gamete (see Figure 5.1) and, if the genes were unlinked and assorting independently, there would be a 1:1:1:1 ratio among the progeny phenotypes. However, A and B are on the same chromosome and linked and so we expect an excess of A B and a b **parental type** gametes; since the chromosome is the unit of segregation, a B and A b **non-parental type** or **recombinant** gametes are *only* formed when there has been a crossover between the A and B loci.

The frequency of recombination is expressed as the proportion (recombination fraction) or percentage (recombination percentage) of

$$\frac{\text{number of recombinant type gametes}}{\text{total number of gametes}}$$

Note that every crossover results in two recombinant and two parental-type gametes, so that, even if there is always a crossover between A and B, the frequency of recombinant-type gametes is still 50%, a figure that can *never* be exceeded. However, if the loci are extremely close, then crossing-over will rarely occur between them and the percentage recombination will approach zero. Thus, percentage recombination can have any value between 50% (independent assortment) and zero (complete linkage).

These recombination values are approximately additive, so that a simple **linkage map**, showing the sequence of the loci along the chromosome and their relative distances apart in terms of percentage recombination units, can be constructed from the data of several independent two-factor crosses.

For example, in *Drosophila melanogaster* ebony body (e), striped body (sr) and maroon eyes (ma) are all located on chromosome III and two-factor crosses produced the following percentage recombination values:

Cross 1 ma $sr/++$ × ma sr/ma sr % recombination $ma–sr$ = 12
Cross 2 sr $e/+$ $+$ × sr e/sr e % recombination $sr–e$ = 8
Cross 3 ma $e/+$ $+$ × ma e/ma e % recombination $ma–e$ = 21

(note that in all future genotypic formulae the more usual convention will be adopted and wild type alleles indicated just by a +).

The only way a linkage map can be drawn so that the distances are approximately additive is

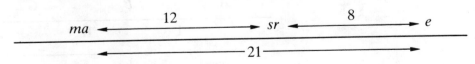

These distances are not precisely additive, because (1) ma and e are the furthest apart and, therefore, the map distance between them calculated by recombination frequency will be the least accurate, and (2) the map distances are calculated from three different sets of experimental data and so are subject to different sampling errors.

Although this is the simplest method for constructing a linkage map, it is more usual to analyse the data from a **three-point test cross**, particularly when the loci are very closely linked. In a three-point backcross one parent is heterozygous for three pairs of linked genes and the crosses are of the type

$$\frac{ma \; sr \; e}{+ \; + \; +} \times \frac{ma \; sr \; e}{ma \; sr \; e} \quad \text{OR} \quad \frac{ma \; + \; +}{+ \; sr \; e} \times \frac{ma \; sr \; e}{ma \; sr \; e}$$

The analysis of three-point backcross data is described in Section 5.4. Look carefully and you will see that this analysis provides rather more information than the sum of the three separate two-point crosses.

5.3 Multiple Crossing-over

In most organisms there are usually two, three or more crossovers along the length of each bivalent, so that sometimes more than one crossover will occur within a particular genetic interval. However, these multiple exchanges are relatively infrequent, as if p is the chance of a crossover occurring between two loci, then the chance of two or three crossovers occurring in that interval is p^2 or p^3 (this assumes that cross-overs occur independently of each other, something that is not always true).

Double crossing-over occurs when there are two separate exchanges, and the three possible types, distinguished by which pairs of non-sister chromatids are involved in each exchange, are shown in Figure 5.2.

Double crossing-over can only be detected when at least three pairs of linked genes are segregating. For example, if two-strand double crossing-over occurred between A and B, the crossovers would cancel out each other and all four chromatids would appear to be non-recombinant.

If we know the recombination fractions between different pairs of markers, we can predict the expected proportions of the different progeny phenotypes in a cross.

Consider the backcross

$$\frac{A \; B \; C}{a \; b \; c} \times \frac{a \; b \; c}{a \; b \; c}$$

In this example it is only necessary to calculate the proportions of the eight types of gamete produced by the heterozygous parent, as it is the genotypes of these gametes which determine the phenotypes of the progeny.

A	B	C
I		II
a	b	c

Let I and II be the intervals between A and B and between B and C. Let p be the recombination fraction for interval I (i.e. the probability of recombination occurring between A and B; note that a recombination fraction of 0.1 is the same as 10% recombination) and let q be the recombination fraction for II.

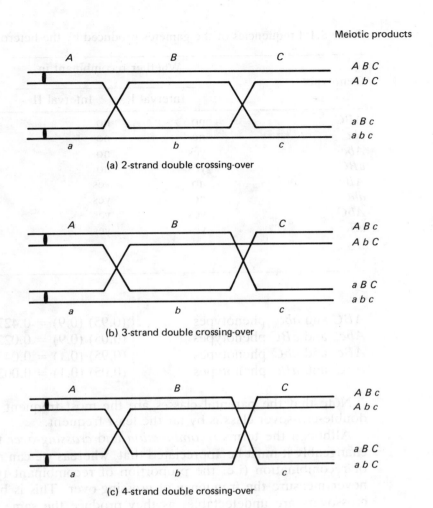

Figure 5.2 *Double crossing-over. The three types of double crossing-over are distinguished by which pairs of non-sister chromatids are involved in each crossover. Each leads to a different combination of genotypes among the meiotic products. Note that if we consider only the segregation of the A and C loci, the two-strand double crossover yields four parental type gametes — in effect, the crossovers have cancelled each other out*

It is important to remember that (1) if p is the chance of recombination occurring in I, then $(1 - p)$ is the chance of no recombination occurring in that interval, and (2) the probability of two independent events occurring simultaneously is the product of their independent probabilities. Thus, the proportion of *ABC* and *abc* gametes, which are non-recombinant in both I and II, is $(1 - p)(1 - q)$ and each of these gametic types occurs with a frequency of $\frac{1}{2}(1 - p)(1 - q)$.

The eight different gametes and their relative frequencies are set out in Table 5.1

Thus if there is 5% recombination in I ($p = 0.05$) and 10% in II ($q = 0.1$), then the expected proportions of the progeny phenotypes will be

Table 5.1 Frequencies of the gametes produced by the heterozygous *ABC/abc* parent

| Genotype | Whether recombinant in | | Frequency |
	Interval I	Interval II	
ABC	no	no	$\frac{1}{2}(1-p)(1-q)$
abc	no	no	$\frac{1}{2}(1-p)(1-q)$
Abc	yes	no	$\frac{1}{2}(p)(1-q)$
aBC	yes	no	$\frac{1}{2}(p)(1-q)$
ABc	no	yes	$\frac{1}{2}(1-p)(q)$
abC	no	yes	$\frac{1}{2}(1-p)(q)$
AbC	yes	yes	$\frac{1}{2}(p)(q)$
aBc	yes	yes	$\frac{1}{2}(p)(q)$
			1

ABC and *abc* phenotypes	$\frac{1}{2}$ (0.95) (0.9) = 0.4275 each
Abc and *aBC* phenotypes	$\frac{1}{2}$ (0.05) (0.9) = 0.0225 each
ABc and *abC* phenotypes	$\frac{1}{2}$ (0.95) (0.1) = 0.0475 each
AbC and *aBc* phenotypes	$\frac{1}{2}$ (0.05) (0.1) = 0.0025 each

Note that the parental classes are the most frequent and, as expected, the double-crossover class is by far the least frequent.

Although the terms *recombination* and *crossing-over* tend to be used interchangeably it must be appreciated that, whereas we can measure the frequency of recombination (i.e. the proportion of recombinant type gametes), we can never measure the frequency of crossing-over. This is because many multiple crossovers are undetectable, as they produce the same gametic genotypes as either no crossover or a single crossover. Only in organisms where the products of individual meioses can be studied (see Section 5.5) can we even distinguish between some of the multiple-crossover events.

5.4 The Three-point Test Cross

Mapping by the use of a three-point test cross is best illustrated by analysing a set of backcross data. The following progeny phenotypes were obtained in a backcross using *D. melanogaster* (*cn*, cinnabar eyes; *b*, black body; *vg*, vestigial wings) and you are asked to construct the relevant linkage map:

+ + +	92	+ + *vg*	6	
cn b vg	70	*cn b* +	9	
cn + *vg*	792	*cn* + +	86	Total 2000
+ *b* +	868	+ *b vg*	77	

Since this is a backcross, one parent must be the triple recessive homozygote and the other parent must be heterozygous at each of the three loci; furthermore, the progeny phenotypes correspond exactly to the genotypes of the gametes produced by the heterozygous parent. But we do NOT know how these genes are arranged in the heterozygous parent, or the order of the loci along the chromosome. However, so long as all three loci are fairly closely linked, the two parental gametic combinations will be indicated by the most frequent progeny phenotypes: these are $cn + vg$ and $+ b +$ (note they are always reciprocally related). However, remember that this distinction becomes less clear as the degree of linkage decreases and as the number of loci segregating increases: nevertheless, if we do make an incorrect deduction, we will obtain a percentage recombination value in excess of 50%, and we know that this can not be correct.

In this cross, as is often possible, we can determine the order of the loci along the chromosome by inspection. To do this it is necessary to identify the double-crossover classes and, usually, these are the least frequent classes: if recombination occurs in both of the possible map intervals, the result is the exclusive exchange of the central marker. Look at Figure 5.2(a): the parental combinations are ABC and abc and double recombination has produced AbC and aBc recombinants — only the central markers, B and b, have been exchanged. In this example the double recombinants are $+ + vg$ and $cn\ b +$ and the parental combinations have been identified as $cn + vg$ and $+ b +$; to produce the double recombinants only cn and $+$ need to be exchanged, so that the cn locus must be located between b and vg. The cross can now be correctly written

$$\frac{+\ cn\ vg}{b + +} \times \frac{b\ cn\ vg}{b\ cn\ vg}$$

Note that it is not essential to determine the order of the loci in this way; the order will become apparent as soon as the percentage recombination values are calculated.

Next we calculate the percentage recombination between *each pair* of loci; in each calculation we ignore the third marker segregating and proceed just as in a two-point test cross.

Thus, the percentage recombination between b and vg

$$= \frac{92 + 70 + 86 + 77}{2000} \times 100 \qquad = 16.25\%$$

the percentage recombination between cn and vg

$$= \frac{6 + 9 + 86 + 77}{2000} \times 100 \qquad = 8.9\%$$

and the percentage recombination between b and cn

$$= \frac{92 + 70 + 6 + 9}{2000} \times 100 \qquad 8.85\%$$

Since the map distances are approximately additive, the only possible gene order is

$$b \longleftarrow\!\!\!-8.85\!-\!\!\longrightarrow cn \longleftarrow\!\!\!-8.9\!-\!\!\longrightarrow vg$$
$$\longleftarrow\!\!\!\!-16.25\!-\!\!\!\!\longrightarrow$$

In this example each of the three map distances has been calculated from the same data and therefore we expect the map distances to be exactly additive. They are clearly not so (8.85 + 8.9 ≠ 16.25), and the reason is that we have *underestimated* the percentage recombination between *b* and *vg*. The *b cn* + (9) and + + *vg* (6) gametes are the result of recombination occurring both between *b* and *cn* and between *cn* and *vg*:

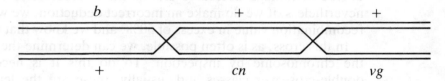

yet, when we calculated the percentage recombination between *b* and *vg*, each of these 15 gametes was scored as non-recombinant, although they had been scored as recombinant between *b* and *cn* and between *cn* and *vg*. Hence, the percentage recombination between *b* and *vg* has been underestimated by *twice* the frequency of the double crossover gametes, i.e. by

$$\frac{2 \times (6 + 9)}{2000} \times 100 = 1.15\%$$

and the correct percentage recombination between *b* and *vg* is 16.25 + 1.15 = 17.75%.

Further points to note in solving linkage problems are:

(1) In the foregoing example you were not told the parental genotypes or how the genes were arranged in the heterozygous parent: in most linkage problems this information is wholly or partly provided and you only have to construct the linkage map.

(2) If you are analysing data from a three- or four-factor cross and you suspect that linkage is involved, you can start by assuming that all the loci are linked. If it turns out that two of the loci are unlinked, you will find that there is 50% recombination between them, i.e. they are assorting independently.

(3) One or more of the loci may be sex-linked, in which event the data will be tabulated separately for the males and females.

(4) Although loci on non-homologous chromosomes always assort independently, two loci on the same chromosome will also be unlinked if they are so far apart that recombination always occurs between them.

(5) The map distance between two loci can be expressed in several different ways. Thus, 1% recombination or a recombination fraction of 0.01 can also be expressed as 1 map unit (m.u.), 1 recombination unit (r.u.) or 1 centimorgan (cM).

(6) In *Drosophila melanogaster* (but not in other species of *Drosophila* or in other organisms) there is NO recombination in males; thus, males only produce parental-type gametes.

5.5 Tetrad Analysis

Many lower eucaryotes, including *Chlamydomonas* and yeast, are haploid for most of their life cycle and the diploid phase is largely restricted to the zygote formed by the fusion of two haploid cells from different strains. This zygote rapidly undergoes meiosis and the four meiotic products remain together as a **tetrad**; this enables special types of genetic analysis not possible in higher eucaryotes, where only the random products of many separate meioses can be analysed.

Tetrad analysis is a complex topic, but since it is so important in certain types of genetic analysis, particularly the understanding of recombination at a molecular level, a very basic description is given here.

In *Saccharomyces cerevisiae*, bakers' yeast, the four haploid meiotic products, the **ascospores**, are held together within a spherical sac-like structure, the **ascus** (Figure 5.3). Let us consider a cross between two haploid parental strains *AB* and *ab*. The diploid zygote will be *AaBb* and meiosis will produce three (and only three) different types of ascus (Figure 5.4); these asci are termed **unordered**, since the spores within them are not in any particular order.

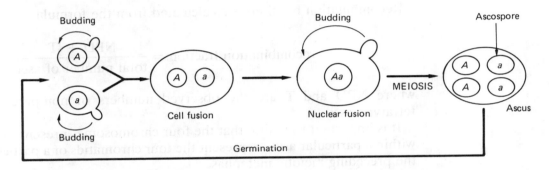

Figure 5.3 *The life cycle of the yeast* Saccharomyces cerevisiae. *Two haploid cells of different genotypes (A and a) fuse to form a binucleate cell, followed by nuclear fusion to produce a* Aa *diploid. The diploid soon undergoes meiosis and produces an ascus containing the four meiotic products, the ascospores. The latter are eventually released and produce new haploid strains. Both the haploid and diploid cells can also reproduce by asexual budding (mitosis)*

If the *A* and *B* loci are unlinked and on different chromosomes, we expect equal numbers of the **parental ditype** (PD, containing only parental type spores) and **non-parental ditype** (NPD, containing all non-parental type spores) asci; this

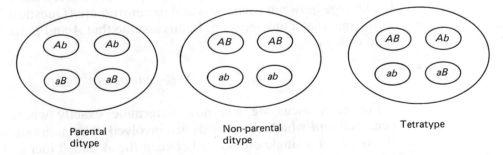

Figure 5.4 *The three types of ascus from the cross* AB × ab. *PD asci occur either in the absence of crossing-over between the* A *and* B *loci or when there has been two-strand double crossing-over. NPD asci only occur as the result of four-strand double crossing-over, while* T *asci are the result of a single crossover*

73

is because, at meiosis, a A chromosome can segregate with either a B or a b chromosome. Hence, *with independent assortment PD = NPD*.

However, if the A and B loci are closely linked (and so on the same chromosome), there will be an excess of the PD asci; this is because the NPD asci are only produced when there has been four-strand double crossing-over between the A and B loci (see Figure 5.2c), so that all four spores are recombinant. Imagine the extreme situation where A and B are so closely linked that recombination rarely occurs between them; in this event the proportion of NPD asci will approach zero and the ratio PD/NPD will approach infinity. Thus, an excess of PD asci indicates linkage, and the greater the departure from unity the greater the degree of linkage. Hence, *with linkage PD > NPD*.

Note that the **tetratype** (T) asci (the result of a single crossover between A and B) tell us nothing about linkage, as they contain two parental type and two recombinant spores, just as with independent assortment.

Recombination fractions are calculated from the formula

$$\text{recombination fraction} = \frac{\text{NPD} + \frac{1}{2}\text{T}}{\text{total number of asci}}$$

where NPD and T are the observed numbers of non-parental ditype and tetratype asci.

It is important to realise that the four chromosomes present in the ascospores within a particular ascus represent the four chromatids of a particular bivalent at the preceding meiotic metaphase.

In *Neurospora crassa*, a filamentous ascomycete, the ascus is linear and **ordered** and contains four pairs of ascospores; meiosis is followed by a single mitosis, so that the tetrad is converted to an octad. Not only are the four spore pairs retained within the ascus, but also they are in a particular order within it. The four spore pairs represent the four meiotic chromatids and are arranged in the same order within the ascus as the chromatids were orientated on the metaphase plate at the preceding meiosis (see Figure 5.5). This makes it possible not only to identify particular chromatids, but also to map centromeres in the same way as gene loci.

Suppose that we cross AB and ab strains and determine the sequence of the spore pairs within a random sample of the asci produced. We can determine whether A and B are linked by using the formula given above and also calculate the linkage between each locus and its centromere (Question 5.7) — this enables us to construct a linkage map. Let us assume that A and B are linked and that the map order is

A–B–centromere

For each ascus we can now determine exactly where crossing-over has occurred *and* which chromatids are involved. The ascus shown in Figure 5.5 is the result of a single crossover between the A and B loci and involving the two non-sister chromatids nearest to the metaphase plate.

Look at the segregation of B and b (ignoring the segregation of A and a). In the ascus both B spore pairs and b spore pairs are side by side; this occurs when there has NOT been a crossover between B and its centromere. This is referred

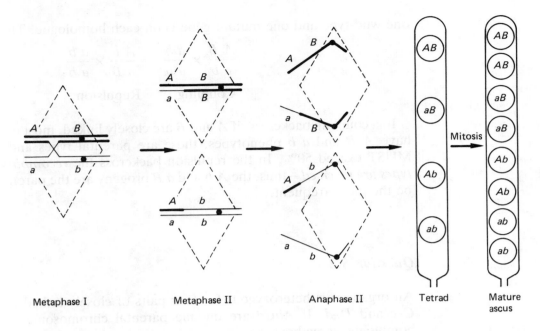

| Metaphase I | Metaphase II | Anaphase II | Tetrad | Mature ascus |

Figure 5.5 *First and second division segregation. First division segregation takes place in the absence of recombination between a locus and its centromere. If a crossover occurs between a locus and its centromere, then second division segregation results*

to as **first division segregation**, because both *B* chromatids separated from both *b* chromatids at the first meiotic division. If you look at the segregation of *A* and *a*, you will see that the *A* and *a* spore pairs are not separated in this way and that the *A* and *a* spore pairs alternate with each other in the ascus; this is because there has been a crossover between the *A* locus and its centromere. This is called **second division segregation**, because the *A* and *a* chromatids do not separate from each other until the second meiotic division.

This makes it a simple matter to calculate recombination fractions between a locus and its centromere. The first and second division segregation asci are scored and the recombination fraction is *one-half of the proportion of second division segregation asci*. This particular recombination fraction can *never* exceed 0.33, because, when a locus and its centromere are assorting independently, two-thirds of the asci will show second division segregation (Question 5.8).

5.6 Questions and Answers

Question 5.1

Distinguish between a coupling and a repulsion backcross. Explain why these two backcrosses do not yield the same proportions of progeny phenotypes.

Answer 5.1

In a **coupling** backcross both of the wild type genes are on one chromosome and both of the mutant genes are on the other homologue. In a **repulsion** backcross one wild-type and one mutant gene is on each homologue. Thus,

$$\frac{A\ B}{a\ b} \times \frac{a\ b}{a\ b} \qquad \frac{A\ b}{a\ B} \times \frac{a\ b}{a\ b}$$

Coupling   Repulsion

In a coupling backcross, if *A* and *B* are closely linked, most of the progeny will have *A B* and *a b* phenotypes; these are parental types and their frequency MUST exceed 50%. In the repulsion backcross the *recombinant and parental types are reversed* – thus, the *A b* and *a B* progeny are the parental types and will be the more frequent.

Question 5.2

An organism is heterozygous for four pairs of closely linked genes, *A–a, B–b, C–c* and *D–d*. If *AbCd* are on one parental chromosome and *aBcD* on its homologue show how crossing over must occur to produce

(a) *ABCd* and
(b) *ABCD* gametes.

Answer 5.2

(a) The two parental chromosomes are

| *A* | *b* | *C* | *d* |

and

| *a* | *B* | *c* | *D* |

To recombine *A* and *B* onto the same chromatid, there must be one crossover between the *A* and *B* loci.

To recombine *C* onto this same chromatid, there must be a further crossover between *B* and *C*:

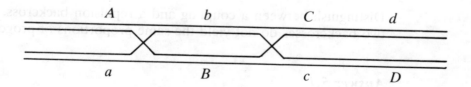

Note that both crossovers must involve the *same* pair of non-sister chromatids — this is known as two-strand double crossing-over.

(b) In order to recombine *D* onto this *ABCd* chromatid, there must be a further crossover between the *C* and *D* loci, again involving the same pair of chromatids:

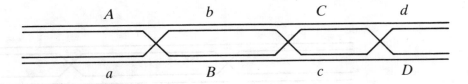

Question 5.3

In an experiment with *Drosophila melanogaster*, females with cut wings (*ct*), vermilion eyes (*v*) and forked bristles (*f*) were mated to wild type males. The F1 females were then backcrossed to *ct v f* females and 1000 progeny were scored:

phenotype		phenotype	
+ + +	341	*ct v* +	96
ct v f	329	+ + *f*	104
ct + +	47	*ct* + *f*	16
+ *v f*	53	+ *v* +	14

Determine whether the loci are linked and, if so, their order along the chromosome and the distance between them. Diagram the cross.

Answer 5.3

Let us assume that the loci are linked and calculate the % recombination between each of the three pairs of markers. Note that the *ct v f* and + + + classes are by far the largest; this is strongly suggestive of linkage and identifies the parental types.

(1) % recombination between *ct* and *v*
Ignore the segregation of *f* and +. All the *ct* + and + *v* progeny are recombinant, i.e. 47 + 53 + 16 + 14 = 130 and % recombination = 130/1000 × 100 = 13%.
   (2) % recombination between *v* and *f*
Ignore the segregation of *ct* and +. The recombinants are *v* + and + *f*, i.e. 96 + 104 + 16 + 14 = 230 and % recombination = 230/1000 × 100 = 23%.
   (3) % recombination between *ct* and *f*
Ignore the segregation of *v* and +. The recombinants are *ct* + and + *f*, i.e. 47 + 53 + 96 + 104 = 300 and % recombination = 300/1000 × 100 = 30%.

The three loci are all linked and the % recombination values are *ct–v*, 13%; *v–f*, 23%; *ct–f* 30%. The only way these can be ordered so that the distances are approximately additive is

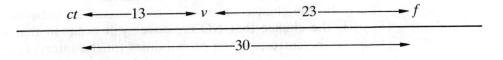

However, $13 + 23 \neq 30$. The reason for this is that the distance between *ct* and *f* has been **underestimated**, because we have ignored the double-crossover classes $+ v +$ and $ct + f$:

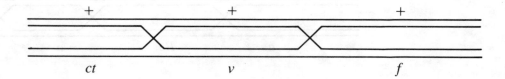

All 30 $ct + f$ and $+ v +$ flies were scored as non-recombinant between *ct* and *f*, whereas there are actually two crossovers in this interval. Hence, the percentage recombination between *ct* and *f* was underestimated by twice the frequency of these classes, i.e. by $2 \times 30/1000 \times 100 = 6\%$, and the corrected map is

$$ct \longleftarrow 13 \longrightarrow v \longleftarrow 23 \longrightarrow f$$
$$\longleftarrow 30 + 6 = 36 \longrightarrow$$

The cross is

P $\dfrac{ct\ v\ f}{ct\ v\ f}$ $\times$ $\dfrac{+\ +\ +}{+\ +\ +}$

F1 $\dfrac{+\ +\ +}{ct\ v\ f}$ $\times$ $\dfrac{ct\ v\ f}{ct\ v\ f}$

Question 5.4

In *Drosophila melanogaster* black body (*b*), purple eye (*pr*) and vestigial wings (*vg*) are closely linked on chromosome 2 in the order *b–pr–vg* and there is 6% recombination between *b* and *pr* and 12% recombination between *pr* and *vg*. What proportion of progeny phenotypes are expected from the cross

$$\frac{b\ +\ vg}{+\ pr\ +} \times \frac{b\ pr\ vg}{b\ pr\ vg} ?$$

Answer 5.4

The male parent is homozygous recessive at each locus and the progeny phenotypes will depend upon the genotype of the female gamete. Crossing-over only need be considered in the female parent.

Let p be the chance of crossing over occurring between *b* and *pr* and thus $(1-p)$ is the chance that NO crossover will occur in that interval. Let q and $(1-q)$ be the corresponding probabilities for the interval *pr–vg*.

$b + vg$ and $+ pr +$ are parental types and there is no recombination in either interval and the overall probability for these two types

$$= (1-p)(1-q) = (1 - 0.06)(1 - 0.12) = (0.94)(0.88) = 0.827$$

$b\ pr +$ and $+ + vg$ are reciprocal recombinant types; they are recombinant between b and pr but non-recombinant between pr and vg.

Thus, the overall probability for these types $= p(1-q) = (0.06)(0.88) = 0.053$.

$b + +$ and $+ pr\ vg$ reciprocal recombinants are only recombinant between pr and vg and the overall probability for these types $= (1-p)q = (0.94)(0.12) = 0.113$.

$b\ pr\ vg$ and $+ + +$ are recombinant in *both* regions and their overall probability $= pq = (0.06)(0.12) = 0.007$.

Thus, the relative frequencies of the eight progeny phenotypes are

$$
\left\{
\begin{array}{ll}
b + vg & 0.413 \\
+ pr + & 0.413
\end{array}
\right.
\quad \text{(i.e. one-half of 0.827)}
$$

$b + vg$	0.413 (i.e. one-half of 0.827)
$+ pr +$	0.413
$b\ pr +$	0.027
$+ + vg$	0.027
$b + +$	0.056
$+ pr\ vg$	0.056
$b\ pr\ vg$	0.003
$+ + +$	0.003
	1.001

Question 5.5

In *Zea mays* (maize) plants homozygous for the recessive genes f (fine stripe), b (colour booster) and v (virescent seedlings) were crossed with wild type plants and the F1 progeny backcrossed to the homozygous recessive parents. 1000 progeny plants were scored with the following results:

(1) wild type	201
(2) boosted-colour	45
(3) virescent	46
(4) fine-striped	211
(5) boosted and virescent	202
(6) boosted and fine-striped	41
(7) virescent, fine-striped	48
(8) fine-striped, boosted, virescent	206

Analyse these results.

Answer 5.5

The contributions of the homozygous recessive parents (*f b v*) can be ignored and the progeny phenotypes indicate the genotypes of the gametes from the heterozygous parent. There is not 1:1:1:1 segregation, so linkage is possibly involved. Rewriting the results and bracketing together the reciprocally related genotypes:

Genotype of gamete and phenotype of progeny	Class	Observed
$\begin{cases} + + + \\ f\,b\,v \end{cases}$	(1) (8)	201 206
$\begin{cases} f + v \\ + b + \end{cases}$	(7) (2)	47 45
$\begin{cases} + + v \\ f\,b + \end{cases}$	(3) (6)	46 41
$\begin{cases} f + + \\ + b\,v \end{cases}$	(4) (5)	211 202

Now calculate the percentage recombination between each pair of loci:

(i) between *b* and *v*
   Number of recombinants   = 48 + 45 + 46 + 41   =   180
   % recombination   = 180/1000 × 100   =   18%

(ii) between *b* and *f*
   Number of recombinants   = 48 + 45 + 211 + 202   =   506
   % recombination   = 506/1000 × 100   =   50.6%

(iii) between *f* and *v*
   Number of recombinants   = 46 + 41 + 211 + 202   =   500
   % recombination   = 500/1000 × 100   =   50%

Thus, *b* and *v* are 18 map units apart and must be on the same chromosome. *b–f* and *f–v* both show about 50% recombination and so are segregating independently. Either *f* is on the same chromosome as *b* and *v* but so far away from them that it segregates independently or it is on a different chromosome. The cross can be represented

P $\dfrac{f\,b\,v}{f\,b\,v}$ × $\dfrac{+++}{+++}$

F1 $\dfrac{+}{f}\ \dfrac{+\ +}{b\ v}$ × $\dfrac{f\,b\,v}{f\,b\,v}$

NOTES

1. The fraction bar separating the contributions of the male and female parents is broken between *f* and *b–v*, indicating that the *f* and *b–v* loci are unlinked.

2. The genotype of the F1 heterozygous parent could also be written $+/f$, $+ +/bv$.

3. This problem clearly shows that the parental genotypes are not always identified by the largest classes.

Question 5.6

Indicate briefly the special uses of ordered tetrads in genetic analysis.

Answer 5.6

(1) In individual tetrads every heterozygous marker segregates 2:2, confirming Mendel's Principle of Segregation.
(2) Linkage values can be calculated from the relative frequencies of the different classes of tetrad.
(3) It is possible to measure linkage between a marker and its centromere.
(4) It demonstrates that recombination is a precisely reciprocal process. In individual tetrads every *AB* recombinant is matched by a reciprocal *ab* recombinant.
(5) The arrangement of the different spore genotypes in an ordered tetrad enables particular chromatids to be identified by genetic methods.
(6) The type of analysis referred to in (5) shows that each recombination event involves only two of the four meiotic products (i.e. chromatids) although multiple crossovers can involve two, three or all four chromatids. This shows that crossing-over must occur at the four-strand stage of meiosis.

Question 5.7

Derive the formulae used in tetrad analysis for calculating the percentage recombination between **(a)** two linked loci and **(b)** a locus and its centromere.

Answer 5.7

(a) There are only three possible types of ascus and each parental ditype ascus contains **NO** non-parental type spores, each non-parental ditype ascus contains **four** non-parental type spores, each tetratype ascus contains **two** non-recombinant type spores.
The percentage recombination = number of recombinant chromatids/total number of chromatids and, since each spore is derived from one meiotic chromatid,

$$\% \text{ recombination} = \frac{4NPD + 2T}{4(PD + NPD + T)} \times 100 = \frac{NPD + \frac{1}{2}T}{(PD + NPD + T)} \times 100$$

(b) In a sample of asci from the cross $A \times a$ let N_1 asci show first division segregation and let N_2 show second division segregation, i.e.

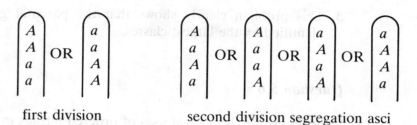

first division   second division segregation asci

(*Note*: only the spore *pairs* are represented)

Each second division segregation ascus is the result of a crossover between A and its centromere:

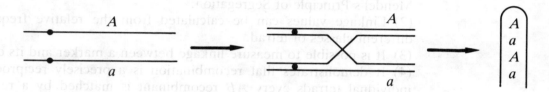

$$\% \text{ recombination} = \frac{\text{number of recombinant chromatids}}{\text{total number of chromatids}} \times 100$$

$$= \frac{2N_2}{4(N_1 + N_2)} \times 100 = \frac{\frac{1}{2} N_2}{N_1 + N_2} \times 100$$

= one-half of the percentage of second division segregation asci.

Question 5.8

Explain why 33.3% recombination between a locus and its centromere is the same as independent assortment.

Answer 5.8

In a Aa diploid each meiosis produces two A and two a spore pairs and, in *Neurospora*, these have to be distributed among the four linearly ordered positions within the ascus. Call these positions 1 to 4 and assume that a A spore pair segregates into position 1:

1		A
2	→	2
3		3
4		4

There are now one *A* and two *a* spore pairs remaining to fill positions 2, 3 and 4. Hence, there is a one-third chance that position 2 will be filled by a *A* spore pair, producing a first division segregation ascus and a two-thirds chance that a *a* spore pair will segregate into position 2 and produce a second division segregation ascus: how the two remaining spores segregate into positions 3 and 4 has no effect on whether the ascus shows first or second division segregation.

Hence, if there is independent assortment between *A* and its centromere, there is a two-thirds chance that the second position will be filled by a *a* spore pair and there will be $66\frac{2}{3}\%$ second division segregation asci. Thus, with independent assortment there will be $33\frac{1}{3}\%$ 'recombination' between a locus and its centromere.

Question 5.9

In *Neurospora crassa* the allelic pairs *A, a* and *V, v* control mating type and normal growth versus visible slow growth. In the cross *A v* × *a V* Howe (1956) recovered the following asci:

type	1	2	3	4	5	6	7	
	Av	*AV*	*Av*	*AV*	*Av*	*AV*	*Av*	
	Av	*AV*	*AV*	*aV*	*aV*	*av*	*aV*	
	aV	*av*	*av*	*Av*	*Av*	*AV*	*AV*	
	aV	*av*	*aV*	*av*	*aV*	*av*	*av*	
asci observed	888	1	126	128	5	3	10	(total 1161)

Are the loci linked and, if so, what are their positions in relation to each other and to the centromere? Draw diagrams to show how crossing over must have occurred to produce type 2 and type 5 asci.

NOTES

1 Calculate the percentage recombination between each marker and its centromere.

2 Calculate the percentage recombination between the *A* and *V* loci.

3 Draw your linkage map so the distances are approximately additive.

4 Draw your crossover diagrams; this cannot be done until you have constructed a linkage map showing the position of the centromere.

Answer 5.9

1 % recombination between *A* and its centromere

Note that asci types 4, 5, 6 and 7 all show second division segregation for *A* and *a*. Thus,

$$\% \text{ recombination} = \frac{\frac{1}{2}(128 + 5 + 3 + 10)}{1161} \times 100 = 6.29\%$$

% recombination between *V* and its centromere

Asci of types 3, 5, 6 and 7 show second division segregation for *V* and *v*. Thus,

$$\% \text{ recombination} = \frac{\frac{1}{2}(126 + 5 + 3 + 10)}{1161} \times 100 = 6.20\%$$

2 % recombination between *A* and *V*

Asci of types 2 and 6 are non-parental ditype (NPD). Asci of types 3, 4 and 7 are tetratype (T). Thus,

$$\% \text{ recombination} = \frac{\frac{1}{2}(126 + 128 + 10) + (1 + 5)}{1161} \times 100 = 11.71\%$$

3 *A* and *V* are clearly linked, and the only way the linkage map can be drawn so that the distances are approximately additive is

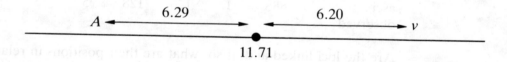

4 The crossover diagrams are:

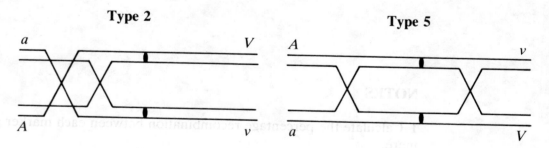

NOTE

The apparent discrepancy in the linkage map (i.e. 6.29 + 6.2 ≠ 11.71) arises because some asci (types 2, 5, 6 and 7) are the result of *two* crossovers and this has led to some errors in scoring the asci. For example, the type 2 asci show first division segregation for *A, a* and were scored as non-recombinant between *A* and the centromere — in fact, there is four-strand double crossing-over in this interval. Similarly, the type 5 asci were scored as non-recombinant between A and v, whereas there are actually two crossovers (two-strand double) in this interval.

5.7 Supplementary Questions

5.1 (a) Define linkage. (b) If the chromosome is the unit of segregation, why do not all the genes along it always remain associated?

5.2 If there is *always* crossing-over between two linked genes, why is it that only 50% of the daughter chromosomes are recombinant?

5.3 In the coupling backcross $a^+b^+/ab \times ab/ab$ you find that 80% of the progeny are a^+b^+/ab and ab/ab. Your interpretation is that a and b are linked with 20% recombination between them. How could you rule out the alternative explanation that the a and b loci are assorting independently and that the a^+b and ab^+ gametes have a greatly reduced (20%) viability?

5.4 (a) What is a linkage group? (b) How many linkage groups are there in *Drosophila melanogaster*?

5.5 What do you understand by 'mating type'? How many types of yeast cell are there?

5.6 In *Neurospora* what events must occur to produce second-division segregation?

5.7 Suggest how you might detect second-division segregation in an organism like yeast, which has an unordered tetrad.

5.8 In *Drosophila* the chromosomes associate in pairs at *mitosis* and crossing-over can occur. Female flies heterozygous for the X-linked alleles y (yellow body) and sn (singed bristles) have wild-type body colour and bristles but occasionally twin spots are found; these are adjacent patches of yellow and of singed tissue. Given that sn is located between y and the centromere, account for these twin spots in terms of mitotic crossing-over.

5.9 What is the difference between mitotic crossing-over and sister chromatid exchange?

5.10 Four genes are located on chromosome 2 of *D. melanogaster* as follows:

al	*dp*		*pr*	*c*
0	13		54.5	75.5
aristaless antennae	dumpy wings		purple eyes	curved wings

The map distances are expressed in recombination units and the position of each locus is measured from the marker nearest the end of one of the chromosome arms. Explain (a) how you can have map distances in excess of 50 r.u. and (b) what this means in practical terms.

5.11 A woman is heterozygous for the X-linked recessive alleles haemophilia (*h*) and colour blindness (*c*). Three of her sons were both colour blind and haemophiliacs, one was just colour blind and three were normal; her three daughters, her parents and her grandparents were all unaffected. Construct a pedigree for this family and give, as far as possible, the inferred genotypes of each individual. Estimate the frequency of recombination between *h* and *c* in this family.

6 The genetics of bacteria and phages

6.1 Bacteria

Much of our knowledge of molecular genetics, including genetic fine structure, mechanisms of recombination, the machinery of DNA replication and protein synthesis and, more recently, the development of recombinant DNA technology and gene cloning, has come from experimental studies using bacteria and their viruses (phages).

The most extensively studied bacteria are *Escherichia coli*, the common gut bacterium, a normally harmless component of the intestinal flora of both man and animals, and the closely related species *Salmonella typhimurium*, the most common single cause of food poisoning in man and animals.

E. coli, frequently referred to in this book, is a procaryote, and, unlike eucaryotes, does not have a nuclear membrane and lacks certain organelles such as mitochondria and chloroplasts; it does not form a spindle at cell division and it reproduces asexually by fission. It is a rod-shaped cell, about 2 μm long and 0.7 μm in diameter, containing a single circular molecule of DNA, the bacterial chromosome. This chromosome is about 3800 kb long and over 1100 genes have been identified and mapped on it.

(a) Genetic Nomenclature in Bacteria

In *E. coli* (and in most bacteria) mutant genes are symbolised by a three-letter lower case italicised abbreviation or acronym of the phenotype:

leu   leucine requirement
lac   inability to ferment lactose
ton   resistance to phage T1 (*T-one*)

When more than one gene affects the same character, each is distinguished by adding a capital letter to the symbol: *trpA*, *trpB* and *trpC* are three of the genes in the *trp* operon.

Finally, separately isolated mutants are distinguished by adding a number:

lac-15 is the 15th isolated mutant in the *lac* operon
trpA23 indicates that the 23rd isolated *trp* mutation occurred within the *trpA* gene.

Where there may be any ambiguity, superscript ($+$) and ($-$) signs can be added to the symbols to designate wild type and mutant genes. However, where there is no confusion, it is usual to refer to the *lacZ* gene and the *lacZ* gene product and to omit the ($+$) superscript.

Phenotypes are represented by a similar non-italicised symbol; thus, Lac$^+$ and Lac$^-$ describe strains able and unable to ferment lactose.

6.2   Bacteriophages

(a)   Virulent Phages

Bacteriophages, or phages, as they are usually known, are parasites which can only replicate within an infected bacterial cell, and to do this they use, to varying degrees, components of the replicational, transcriptional and translational machinery of the host cell.

One of the best-known phages is T4, a large phage (Figure 6.1) which infects *E. coli*. It is a virulent phage, and whenever it infects a sensitive bacterium, it enters the lytic cycle:

(1) A particle of T4 attaches by its tail fibres to a receptor site on the bacterial cell membrane.
(2) The contractile sheath contracts.
(3) The rigid tail core penetrates the cell membrane rather like a hypodermic needle.
(4) The phage DNA is injected into the infected cell and takes over the biosynthetic machinery of the cell — more phage chromosomes and protein coat components are synthesised.
(5) These components are assembled into mature phage particles.
(6) 30 min after infection the cell wall breaks open (or **lyses**), releasing 200–300 infectious phage particles.

The T4 chromosome is a linear molecule of DNA about 166 kb long; over 150 genes have been identified and mapped.

(b)   Temperate Phages

Temperate phages, when they infect a host cell, either can enter the lytic cycle just like a virulent phage or may **lysogenise** the host cell. In a **lysogenic cell** or **lysogen** a complete phage genome is present but it is *not active*; in this dormant state the phage genome is known as **prophage**.

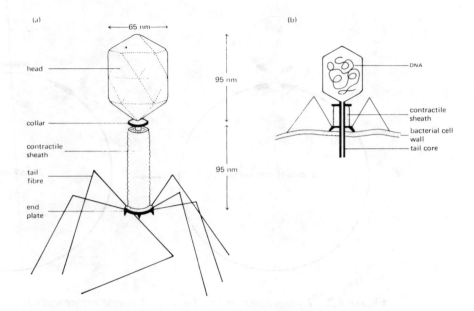

Figure 6.1 *Bacteriophage T4.*

(a) Phage T4 consists of a protein head, made up of many identical protein sub-units and containing a single molecule of DNA, and a complex tail. The protein tail consists of a rigid hollow core, a contractile sheath and a spiked end plate with six attached tail fibres.

(b) Upon infection of E. coli, *the phage attaches by its tail spikes and fibres to a special receptor site on the bacterial cell wall; the tail sheath contracts, the rigid core is 'pushed' through the cell wall and the DNA is injected into the bacterium*

The best-known temperate phage is lambda (λ). This resembles T4 but it has only a single tail fibre and contains a linear molecule of DNA 48.6 kb long. When λ infects a sensitive cell of *E. coli* (Question 6.4):

(1) The linear molecule of phage DNA is injected into the host cell.

(2) The DNA molecule immediately circularises (Question 3.7).

(3a) The phage genome enters the lytic cycle (as described for T4) and eventually lyses the host cell, releasing about 100 phage progeny.

OR

(3b) A single reciprocal recombination event occurs between specific sequences on the circularised phage chromosome and the circular bacterial chromosome, so that the phage genome is integrated into the continuity of the bacterial chromosome (Figure 6.2). This prophage is in an inactive state and replicates as though it were a part of the bacterial chromosome.

Lysogenic cells have two special properties: (1) they cannot be infected by a second λ phage – they are immune to superinfection; and (2) at some stage in the future the λ prophage can be excised from the bacterial chromosome and enter the lytic cycle—lysogenic cells have the potential to release free phage.

Not all temperate phages integrate into the host chromosome in this way. With phage P1, which also infects *E. coli*, the phage DNA is maintained like a plasmid (Section 6.3a), an independently replicating circular molecule of DNA.

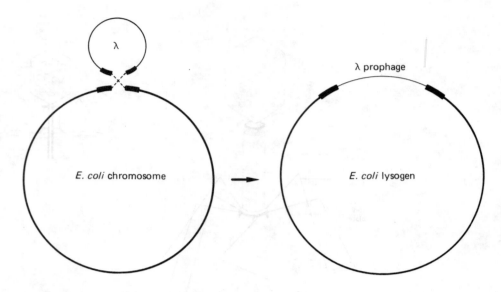

Figure 6.2 *Lysogenisation by phage* λ. *A single reciprocal crossover between specific nucleotide sequences on the* λ *and* E. coli *chromosomes inserts* λ *into the continuity of the bacterial chromosome*

6.3 Genetic Analysis in Bacteria

Bacteria are haploid and do not reproduce sexually, so that it is not possible, as it is in eucaryotes, to construct linkage maps by the analysis of meiotic products. Nevertheless, it is possible to obtain transient partial diploids, or **merozygotes,** containing a complete circular bacterial chromosome *and* a linear segment of a chromosome derived from a different strain (the **donor**); recombinants are formed when crossing-over occurs between this segment and the homologous region of the bacterial chromosome.

In bacteria recombinants are usually rare — sometimes as infrequent as 1 in 10^7 (10^{-7}) parental bacteria — and special methods are necessary to detect them. **Selective techniques**, first developed in mutation studies (Questions 6.1 and 11.1), permit the rare mutants or recombinants to grow and to form colonies on solid medium but do not allow growth of the parental strain.

The simplest **selective technique** uses a mutant resistant to a particular drug (e.g. streptomycin) or phage (e.g. T4). The drug or phage-sensitive parental strain is plated on a fully nutrient solid medium (nutrient agar) containing the drug or sprayed with the phage; the parental strain is killed by the drug or phage and any resistant survivors either are due to mutation from sensitivity to resistance or, if the parental recipient strain was a partial diploid carrying a resistance gene on a chromosome fragment, are the result of crossing-over recombining the resistance gene into the bacterial chromosome.

Three points to remember are: (1) recombination always involves an *even* number of crossovers (Question 6.5), usually two; (2) reciprocally related recombinant bacteria are never recovered; and (3) it is not usually known how many non-recombinant merozygotes are present, so that it is not possible to calculate map distances in terms of percentage recombination.

There are three basic methods for transferring segments of chromosomal DNA from one bacterial cell to another.

(a) Conjugation

Many of the important properties of bacteria — in particular, the resistance of certain strains to a range of antibiotics — are the consequence of the presence of one or more plasmids. **Plasmids** are circular molecules of double-stranded DNA normally existing free in the cell (when they are said to be **autonomous**) and replicating independently of but in harmony with the bacterial chromosome. They are not normally necessary for bacterial growth and survival and, in effect, they are small, non-essential, additional chromosomes.

Conjugation, the most important method of DNA transfer in *E. coli* requires the presence in the donor cell of a particular plasmid known as the **F (fertility) plasmid** or **F factor**. Cells carrying an autonomous F are known as F^+ and cells without F are F^-.

F is a large plasmid, containing about 94.5 kb of DNA, and it has three very important properties.

First, it encodes a flagellum-like structure, the **F pilus**, found on the surface of each F^+ cell. The F pilus is about 8 nm in diameter, is between 2 μm and 20 μm long and has an axial hole the diameter of a duplex molecule of DNA; an F pilus can attach by its tip to a receptor site on the surface of an F^- cell. Thus, when F^+ and F^- cells are mixed, they join together in pairs, or **conjugate**.

Second, F can undergo a special type of replication known as transfer replication (Question 6.7). In transfer replication a copy of F is transferred through a conjugation tube into the F^- cell, converting it to F^+. It is possible that the F pilus is also the conjugation tube but it is more likely that the F pilus simply brings the F^+ and F^- cells into very close contact, enabling the formation of a cytoplasmic bridge between them (Figure 6.3a). In these $F^+ \times F^-$ matings there is little or no transfer of chromosomal DNA.

Third, F contains special sequences which enable it to **integrate** into the bacterial chromosome at any one of a number of different sites (this is the result of a single reciprocal crossover occurring between F and the bacterial chromosome), forming an **Hfr** (high-frequency recombination) cell. In Hfr cells F is still fully functional, so that all Hfr cells have an F pilus, but otherwise the integrated F behaves like a group of bacterial genes.

When Hfr and F^- cells are mixed, then, as in $F^+ \times F^-$ crosses, the F pilus promotes the formation of conjugal pairs but now the integrated F promotes the transfer of a copy of the bacterial chromosome, through a conjugation tube, into the F^- cell. Transfer always commences from a site within F and always proceeds in the same direction (relative to the orientation of F); this means that there is the oriented transfer of a linear segment of the bacterial chromosome into the F^- cell; the transferred segment can now recombine with the chromosome of the F^- recipient (Figure 6.3c).

Transfer of the entire chromosome takes 100 min but, because the mating pairs are in constant Brownian motion, the fragile conjugation tube connecting the Hfr and F^- is spontaneously broken as a function of time. Thus, if the Hfr chromosome is transferred in the sequence

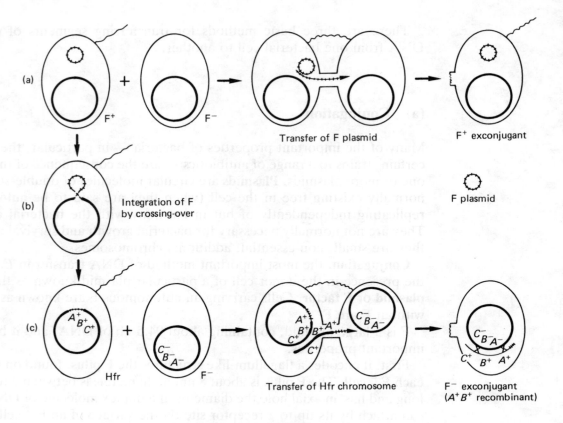

Figure 6.3 *Conjugation in* E. coli.

(a) Mating between an F⁺ and an F⁻ cell.

The F pilus joins the pair of F⁺ and F⁻ cells. A conjugation tube forms and, as a result of transfer replication, a copy of F is transferred into the F⁻ cell: this circularises and converts the F⁻ cell to F⁺.

(b) F integrates into the bacterial chromosome by a single reciprocal crossover forming an Hfr chromosome.

(c) Mating between an Hfr and an F⁻ cell.

Once transfer replication is initiated within F, the chromosomal genes behave as though they were part of F, so that part of F and some of the adjacent chromosomal genes are transferred into the F⁻ cell. In the F⁻ cell recombination integrates various segments of the transferred DNA into the recipient chromosome

$$\overset{\longleftarrow}{A \qquad B \qquad C \qquad D \qquad E \;\text{--}\;\text{--}\;\text{--}\; Y \qquad Z}$$

then nearly every exconjugant F⁻ cell will have received a chromosome segment containing A^+, a smaller number will also receive B^+, still fewer will receive C^+ and so on, i.e.

$$\frac{\qquad A^+ \qquad}{}$$
$$\frac{A^+ \qquad B^+ \qquad}{}$$
$$\frac{A^+ \qquad B^+ \qquad C^+}{}$$

etc.

Once in the F$^-$ cell, every transferred Hfr gene has an equal chance of being recombined into the F$^-$ chromosome. Thus, if we select A^+ recombinants (N.B.: the selected marker must be at the leading or proximal end of the Hfr chromosome) and then test these to determine which other markers they have inherited, we find (for example) that 90% of the A^+ recombinants have also inherited B^+, 80% have inherited C^+, 60% have inherited D^+, and so on. Thus, the location of the markers C^+, D^+ and E^+, etc., is shown by the frequency with which each marker appears among the A^+ recombinants; this is known as mapping by **gradient of transmission**. A refinement of this method is mapping by **interrupted mating**. At varying times after the commencement of mating the Hfr−F$^-$ conjugal pairs are separated by violently agitating the mating mixture; the culture is then diluted to prevent conjugal pairs re-forming and is plated to select A^+ recombinants. These are then tested to determine the proportion inheriting each of the more distal unselected markers. Since conjugal pairs form very rapidly and since the chromosome is transferred at a uniform rate (38 kb/min), a particular marker will always first enter the F$^-$ recipients at a particular time after mating; if mating is interrupted before that time, then that marker will not appear among the A^+ recombinants. This makes it possible to construct a linkage map with the map distances expressed in time units; any unmapped marker is easily assigned a map position by determining the **time at which it first appears among the recombinants**.

These methods are most useful when mapping genes more than 2 min apart on the map; if the genes are very closely linked, then it is better to use a more conventional recombinational analysis.

Hfr strains only transfer about 30% of their chromosome at high frequency, as by 30 min almost every conjugal pair has been separated by spontaneous breakage of the conjugation tube. However, F can integrate at a number of sites around the chromosome, so that many different Hfrs are known, each transferring different and overlapping segments of the Hfr chromosome. Thus, by using the data from several different Hfr strains it is possible to build up a composite map of the *E. coli* chromosome (see Figure 6.9).

These novel methods were developed by Elie Wollman, François Jacob and William Hayes in 1956.

(b) Transduction

In **transduction**, discovered by Norton D. Zinder and Joshua Lederberg in 1951, a temperate phage transfers a segment of a bacterial chromosome from a donor strain to a recipient. In *E. coli* phage P1 is the most commonly used transducing phage.

When phage P1 infects *E. coli* and enters the lytic cycle, the maturing phage particles usually package a newly replicated phage chromosome, but a small proportion (10^{-3}) encapsidate a small fragment of the host chromosome instead; these fragments of host DNA are about 100 kb long (the same size as the P1 chromosome) and include 25–30 known loci (i.e. about 1/40 of the bacterial chromosome). Thus, about $1/40 \times 3/1000$ or 7.5×10^{-5} phage will carry a fragment of the host chromosome carrying a *particular* bacterial gene. These **transducing particles** can adsorb normally onto a sensitive recipient cell and inject the fragment of donor DNA into it; once the fragment of donor DNA is

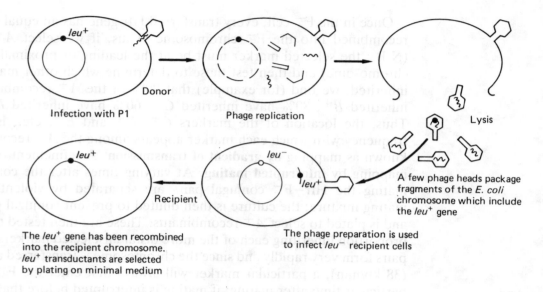

Infection with P1   Donor   Phage replication   Lysis

A few phage heads package fragments of the E. coli chromosome which include the leu⁺ gene

The phage preparation is used to infect leu⁻ recipient cells

The leu⁺ gene has been recombined into the recipient chromosome. leu⁺ transductants are selected by plating on minimal medium

Recipient

Figure 6.4 *P1-mediated transduction in* E. coli

inside the recipient cell, it can recombine with the homologous region of the recipient chromosome and produce a **transductant** (Figure 6.4).

If a suspension of P1 is prepared by lytic growth on a wild type donor strain of *E. coli* and used to infect a population of leucine-requiring (*leu⁻*) recipient cells and the mixture is plated on minimal medium, then after 2–3 days many leucine-independent (*leu⁺*) colonies are found growing on the plates. A few of these may be due to back-mutations of the *leu⁻* allele in the recipient to *leu⁺* but most will be transductants, where the *leu⁺* gene on the fragment of donor DNA has been recombined into the recipient chromosome. About 1 transductant is found for every 10⁵ phage particles used and the selective technique described enables the detection of these rare genetic recombinations. This transduction is written

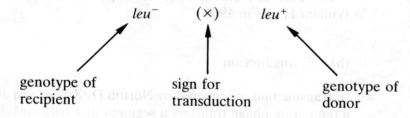

leu⁻   (×)   *leu⁺*

genotype of recipient   sign for transduction   genotype of donor

and represented diagrammatically

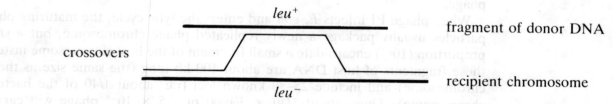

leu⁺

fragment of donor DNA

crossovers

recipient chromosome

leu⁻

Since each fragment carries 25–30 known genes, it is possible to carry out transductions between strains differing by two or more pairs of closely linked genes; this makes it possible to carry out recombinational analysis and to

94

construct linkage maps for small segments of the chromosome. Two methods are most commonly used: **co-transduction** and **three-point transduction**.

(1) *Co-transduction* The closer together two markers are on the donor fragment the less the likelihood of recombination occurring between them and so the greater the probability that they will be simultaneously recombined into the recipient chromosome, i.e. the greater the frequency of co-transduction or joint transduction.

(2) In *three-point transductions* (as in three-point test crosses in higher organisms) the loci are ordered so as to explain the data with the minimum number of crossovers — this is, of course, the basic premise of all recombinational analysis.

Transduction is widely used in mapping a range of bacterial genomes. It is particularly useful in fine structure analysis — for constructing maps showing the order of mutant sites within a gene and the order of genes that are less than 100 kb apart; this is in contrast to conjugational mapping, which is most useful in mapping genes more widely spaced and up to 1000 kb apart.

(c) Transformation

Transformation was the first method of genetic transfer to be discovered in bacteria, and it was through transformation experiments that Avery and his co-workers (Question 2.6) first demonstrated that DNA was the genetic material.

DNA is extracted from a donor strain, purified and applied to a genetically different recipient. The double-strained donor DNA binds to receptor sites on the cell membrane of the recipient cell and enters it. Once in the recipient cell, the fragment of transforming DNA can recombine with the recipient chromosome and generate a **transformant**. Transformation, like transduction, is a rare event, and transformants must be selected by plating on an appropriate medium.

The fragment of transforming DNA is 7–9 kb long and so only includes two or three identified bacterial genes.

(d) F-prime Plasmids

An F plasmid can, by reciprocal recombination, integrate into the bacterial chromosome to form an Hfr cell (Section 6.3a) and, by precisely the reverse process, can excise from it to re-form an F^+ cell. Just occasionally this excision is imprecise and a reciprocal crossover occurs between a site within F and a site on the bacterial chromosome (Figure 6.5). This produces an F plasmid in which a non-essential part of the plasmid has been replaced by several adjacent bacterial genes; these are known as **F-prime** or **F′** plasmids.

Since F′ plasmids are stably inherited, strains carrying them are permanent merozygotes with a complete set of genes on the bacterial chromosome and a partial set on the F′ plasmid. Because there are many different Hfr strains, it is possible to isolate F′ plasmids carrying almost any bacterial gene.

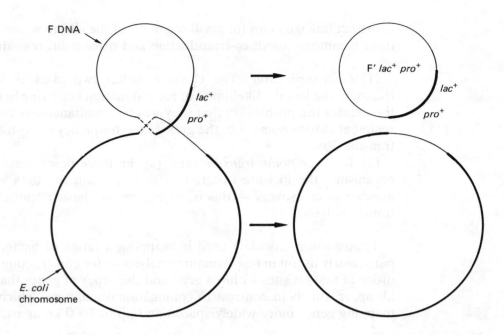

Figure 6.5 *The origin of an F′ lac⁺ pro⁺ plasmid. In this example the F′ plasmid has arisen as the result of a recombination occurring between a site on the bacterial chromosome and a site within F*

These F′ plasmids are particularly useful, as they allow the construction of stable partial diploids and permit complementation tests to be carried out in bacteria (Section 7.6(b) and Questions 10.1 and 10.2).

6.4 Genetic Mapping in Bacteriophages

Individual phages can be recognised because each one causes a **plaque** to form when a dilute suspension is plated onto a lawn of sensitive bacteria (see Question 6.3); this is a clear area within the opaque lawn of confluent bacterial growth caused by the reproduction of the phage. Different phages produce plaques of characteristic size and morphology, and many mutants were first recognised because they produced plaques with an altered morphology. Other mutants, known as **host range mutants**, infect different bacterial hosts from those of the wild type.

One very extensively analysed region of the phage T4 genome includes the *rIIA* and *rIIB* genes (see also Questions 6.9, 9.2, 9.3 and 11.5). These *rII* mutants (*r*apid lysis) form larger plaques with sharper edges than the wild type *r⁺* phage. The *r⁺* phage produce small fluffy plaques on *E. coli* strain B and on *E. coli* strain K, whereas the *rII* mutants form *r* type plaques on *E. coli* B and no plaques at all on *E. coli* K; only wild type *r⁺* phage can grow on *E. coli* K (Figure 6.6).

Note that the *rII* mutants have an altered plaque morphology on *E. coli* B and an altered host range on *E. coli* K.

Suppose that we have isolated a number of *rII* mutants (*r-1 r-2 r-3*, etc.) and wish to construct a linkage map showing the relative positions of these mutant

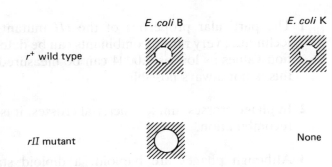

Figure 6.6 *Plaques formed by the wild type and* rII *mutants of T4. The* rII *mutants do not form plaques on* E. coli *K and they form large sharp-edged plaques on* E. coli *B. T4* rII+ *forms small fuzzy plaques on both strains*

sites within the *rII* region. The simplest method is to cross the mutants in all pairwise combinations (*r-1* × *r-2, r-1* × *r-3, r-2* × *r-3*, and so on) and determine the percentage recombination between each pair of mutant sites; the mutant sites are then ordered, as in crosses in higher organisms, so that the map distances are approximately additive. The procedure is as follows:

(1) *E. coli* B is simultaneously infected with two different *rII* mutants (*r-1* and *r-2*); the two different genomes will replicate and undergo genetic recombination.

After the cells have lysed, the phage lysate is harvested; this will include *r-1, r-2, r+* and *r-1 r-2* genomes.

(3) A suitable dilution of the lysate is plated on *E. coli* B. All the phages will reproduce and form plaques. The number of plaques × the dilution factor provides an estimate of the total phage present in the lysate.

(4) A suitable dilution of the lysate is plated on *E. coli* K. ONLY the *r+* recombinants can grow. Note that the reciprocal recombinant class (the *r-1 r-2* double mutant) cannot form plaques and so cannot be detected; thus, the best estimate of the total number of recombinants is **twice** the number of *r+* recombinants.

(5) Repeat the above for the other pairwise combinations (*r-1* × *r-3, r-2* × *r-3*, etc.).

(6) Calculate the percentage recombination frequencies:

$$\% \text{ recombination} = \frac{2 \times (\text{number of } r^+ \text{ phage})}{\text{total number of phage}} \times 100$$

$$= \frac{2 \times (\text{number of plaques on K})}{\text{number of plaques on B}} \times \text{dilution factor} \times 100$$

(7) The mutant sites are ordered so that the map distances are approximately additive (see Question 6.9).

NOTES

1 The particular properties of the *rII* mutants allows the use of a selective technique; very rare recombinants can be detected and percentage recombination values as low as 0.0004 can be measured. In crosses with other mutants this is not always possible.

2 In phage crosses, unlike bacterial crosses, it is possible to measure percentage recombination.

3 Although phages are haploid, a diploid state is effectively produced by simultaneously infecting the host cell with two different phage mutants.

6.5 Questions and Answers

Question 6.1

Why are bacteria so suitable as organisms for genetic study?

Answer 6.1

(1) They can, like *Neurospora*, be grown on a simple chemically defined *minimal* medium containing inorganic salts and a carbon source (frequently glucose or lactose).
(2) They can be grown in pure cultures.
(3) Many different mutants can be easily isolated; these include

> (i) auxotrophs — mutants only able to grow when a particular amino acid, vitamin or nucleotide is added to the medium,
> (ii) mutants unable to ferment a particular sugar such as lactose (*lac⁻*) or arabinose (*ara⁻*),
> (iii) antibiotic resistance mutants,
> (iv) phage resistance mutants,
> (v) mutants unable to carry out genetic recombination.

(4) The use of selective techniques is possible. For example, if you wish to determine how frequently a proline-requiring (*pro⁻*) mutant of *E. coli* undergoes spontaneous mutation back to the wild type, it is only necessary to plate about 10^8 cells of *E. coli pro⁻* on a minimal medium plate; the *pro⁻* bacteria cannot grow and every colony that develops will be the result of a mutation to proline-independence.
(5) They have a very short generation time — 20 min under optimal conditions — so that millions of organisms can be produced in a very short time; a single bacterium can produce over a million progeny in less than 7 h (20 generations).
(6) A wide range of methods is available for genetic analysis and genetic manipulation.

Question 6.2

What are the differences between virulent and temperate bacteriophages?

Answer 6.2

When a virulent phage infects a sensitive bacterial cell, it immediately replicates and produces phage-specific proteins; these are assembled into mature phage particles and, after 20–30 min, the host cell is lysed and several hundred phage are released into the medium.

On the other hand, a temperate phage has two options when it infects a sensitive bacterium. If there is an abundant supply of energy (i.e. when growth conditions are favourable), it can enter the lytic cycle and lyse the host cell in just the same way as a virulent phage. However, if the growth conditions are less favourable, it may lysogenise the host cell. In a lysogenic cell the phage genome (prophage) replicates in harmony with the bacterial chromosome but most of the phage genes are not expressed; however, at some time in the future, when growth conditions are again favourable, the prophage may enter the lytic cycle and produce mature phage.

In lambda lysogens, as in lysogens for many other temperate phages, the phage genome is recombined into the continuity of the bacterial chromosome and behaves as though it were a group of bacterial genes. In other instances, notably P1 lysogens, the prophage is maintained as an autonomous plasmid still replicating in harmony with the bacterial chromosome.

Question 6.3

How can you estimate (assay) the number of phage present in a suspension of T4?

Answer 6.3

Individual phages can be identified because they form plaques. If a drop of a mixture containing 100 or so phages is mixed with very many phage-sensitive bacteria (10^7 or more), added to molten soft agar, allowed to cool, poured over the surface of a nutrient plate and allowed to solidify, the bacteria will grow rapidly and form a lawn of confluent growth over the surface of the plate. However, 100 or so bacterial cells will have been infected by T4 phage and in such cells the phage will replicate and lyse the infected cell, releasing 200–300 mature phages; these phages will infect adjacent bacteria and in turn they will be lysed; this process of lysis and infection will continue until the supply of nutrients in the medium is exhausted and bacterial metabolism ceases. The result is a circular clearing or **plaque** in the lawn of confluent bacterial growth.

Each plaque represents the initial infection of one bacterium by one phage. Thus, the titre of the phage suspension can be estimated if we know (1) the number of plaques on the plate, (2) the volume of the phage suspension used to inoculate the plate and (3) the dilution of the phage suspension used to infect the bacteria.

Question 6.4

Make an annotated diagram showing the life cycle of phage lambda.

Answer 6.4

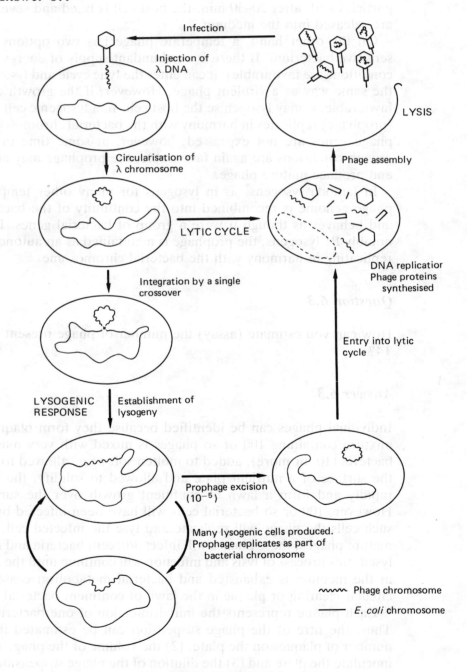

Figure 6.7 *The life cycle of phage λ*

Question 6.5

Explain why, in bacterial recombination, **(a)** two crossovers are necessary to produce a recombinant and **(b)** reciprocal recombinants are never recovered.

Answer 6.5

(a) All bacterial transfer systems produce a merozygote containing a circular recipient cell chromosome and a fragment of donor DNA, and recombination occurs between these two molecules. A single crossover (or any odd number of crossovers) produces an oversize linear genome which would be unable to replicate (Figure 6.8); only an even number of crossovers can maintain the circularity of the recipient chromosome.

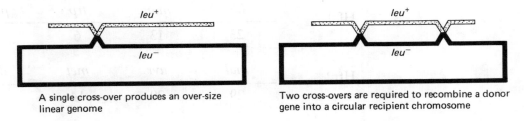

A single cross-over produces an over-size linear genome

Two cross-overs are required to recombine a donor gene into a circular recipient chromosome

Figure 6.8 *Recombination in bacteria involves an even number of crossovers*

(b) First, in the above example the reciprocal recombinant is a fragment of DNA carrying leu^-. This fragment cannot replicate and can never be detected. Second, reciprocal recombinants are never recovered when a selective technique is used to recover one particular class of recombinant. Thus, if A^+ transductants are selected in the transduction $A^- B^- (\times) A^+ B^+$, then $A^+ B^+$ and $A^+ B^-$ recombinants can be detected but never $A^- B^+$ recombinants.

Question 6.6

You have isolated four new Hfr strains of *E. coli* and, by interrupted mating experiments, for each strain determined the markers that are transferred at high frequency and their times of entry into the F^- recipient:

	Hfr1		Hfr2		Hfr3		Hfr4	
markers and times	*man*	13 min	*mal*	29	*lys*	16	*pur*	6
of first entry	*trp*	6	*met*	14	*arg*	9	*trp*	3
	his	23	*thr*	4	*mal*	2	*thr*	31
	pur	3	*uvr*	20	*his*	32	*lac*	23
							gal	14

(*gal, lac, mal, man:* inability to ferment galactose, lactose, maltose and mannitol. *arg, his, lys, met, pur, trp:* requirement for arginine, histidine, lysine, methionine, purines and tryptophan. *uvr:* sensitivity to UV irradiation.)

Any markers not indicated were not transferred at high frequency. Do the data support the concept that the chromosome of *E. coli* is circular? Construct a linkage map of the *E. coli* chromosome with the map distances expressed in minutes; assume that the entire chromosome takes 100 min to be transferred and that the *thr* locus is arbitrarily assigned the values of 0 min and 100 min.

Answer 6.6

(1) Each Hfr transfers a particular group of markers (*c.* 30% of the *E. coli* chromosome) at high frequency; the first marker to be transferred is the marker that first appears among the recombinants, and so on. Thus, the four Hfr strains transfer their chromosomes as follows:

Hfr1	*his*	*man*	*trp*	*pur*
	23	13	6	3

Hfr2	*mal*	*uvr*	*met*	*thr*
	29	20	14	4

Hfr3	*his*	*lys*	*arg*	*mal*
	32	16	9	2

Hfr4	*thr*	*lac*	*gal*	*pur*	*trp*
	31	23	14	6	3

(2) Note that these Hfrs transfer overlapping segments of the genome and that Hfr1 transfers in the opposite orientation to the other three Hfr strains.

(3) Thus, the linkage map can be drawn as in Figure 6.9, supporting the concept that the genome is circular.

(4) The *thr* locus is arbitrarily assigned the position of 0 min and 100 min; *lac* is transferred 8 min after *thr* (data from Hfr4) and so is assigned a map position of 8 min; likewise, *trp* is separated from *thr* by 28 min (data from Hfr4) and is assigned a position of 28 min; *man* is transferred 7 min after *trp* (data from Hfr1) and is assigned a map location of 35 min; and so on.

(5) In Figure 6.9 the small arrows indicate the positions at which F has been integrated into the *E. coli* chromosome. The inner arcs show for each Hfr the group of markers transferred at high frequency and the orientation of transfer.

NOTE

Hfr1 and Hfr4 both transfer *trp* and *pur* but in opposite orientations. In each instance these markers are 3 min apart, confirming that the Hfr chromosome is transferred at the same rate in all strains.

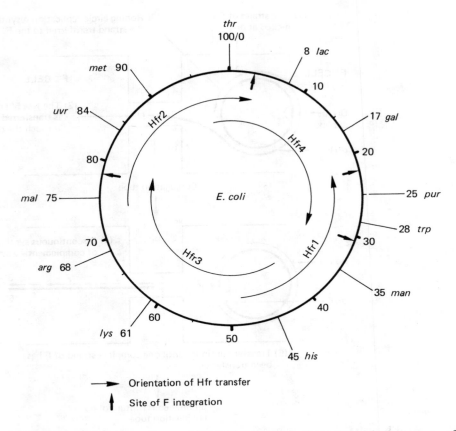

Figure 6.9 *The chromosome of* E. coli. *The inner arcs show the segments of the chromosome transferred at high frequency by four different Hfr strains. The arrowheads show the directions of transfer*

Question 6.7

Using annotated diagrams show how F is transferred from an F⁺ to an F⁻ cell. Briefly state how transfer replication differs from vegetative replication.

Answer 6.7

Transfer replication commences at *oriT* (*ori*gin for *t*ransfer) and proceeds by the rolling-circle mode of replication. Vegetative replication commences at *oriV* (*ori*gin for *v*egetative replication) and proceeds bidirectionally by theta-mode replication (like the *E. coli* chromosome itself).

NOTES

1   In Hfr transfer rolling-circle replication again commences at *oriT* but the transferred strand does NOT become double-stranded; varying segments of

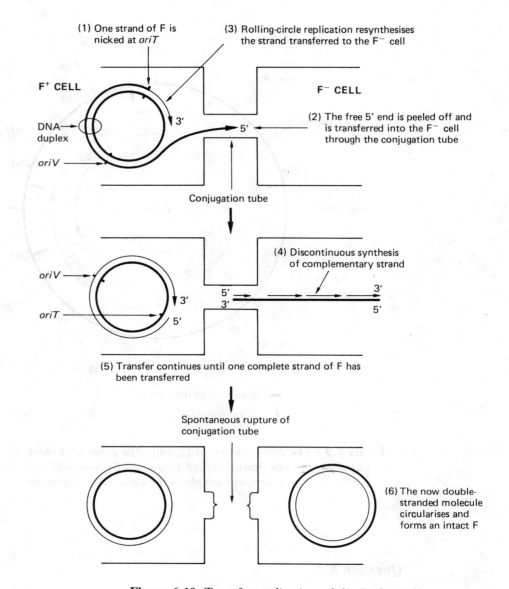

Figure 6.10 *Transfer replication of the F plasmid*

the single strand transferred into the F⁻ cell are recombined into the bacterial chromosome.

2 Since transfer commences *within* F, only part of F is transferred into the F⁻ cell. The recipient exconjugant is, therefore, F⁻.

3 The mode of transfer is such that the bacterial chromosome behaves as though it were part of F (rather than F behaving as part of the bacterial chromosome).

Question 6.8

trp⁻ cys⁻, *trp⁻ pyr⁻* and *pyr⁻ cys⁻* strains of *E. coli* were infected with P1 phage grown on a *trp⁺ pyr⁺ cys⁺* donor. Transductants were selected as shown below and samples of the selected transductants were tested by streaking on minimal medium to determine the proportion that had also inherited the unselected marker from the donor (i.e. the co-transductants):

transduction	selected marker	co-transductants/ total transductants
(i) trp^- cys^- (×) trp^+ cys^+	trp^+	240/400
(ii) trp^- pyr^- (×) trp^+ pyr^+	trp^+	176/370
(iii) pyr^- cys^- (×) pyr^+ cys^+	pyr^+	270/415

(a) For each transduction state the selective medium used.
(b) Deduce the order of the three loci on the *E. coli* chromosome.

Answer 6.8

(a) Transduction (i) and (iii): minimal medium supplemented with cysteine. Transduction (ii): minimal medium supplemented with pyrimidines.

(b) The closer two markers are on the chromosome the higher will be the frequency of co-transduction. The co-transduction frequencies are (i) *trp–cys*, 0.6; (ii) *trp–pyr*, 0.47; and *pyr–cys*, 0.65. From (i) and (ii) the map is either

(A) *trp*   *cys*   *pyr*   (B) *cys*   *trp*   *pyr*

    └—0.60—┘   or   └—0.60—┘└—0.45—┘

    └————0.45————┘

But from (ii) and (iii) it is clear that *pyr* and *cys* (0.65) are closer than *pyr* and *trp* (0.47). Therefore, order must be (A), i.e.

      trp   *cys*   *pyr*

      └—0.60—┘└—0.65—┘

In order to be co-transduced all three loci must lie within a 100 kb segment of the *E. coli* chromosome.

Question 6.9

Three *rII* mutants of T4 are crossed in pairwise combinations and plated at the same dilution on *E. coli* strains B and K; the results were

cross	plaques on *E. coli* B	plaques on *E. coli* K
rIIa × *rIIb*	390	9
rIIb × *rIIc*	460	8
rIIa × *rIIb*	306	13

Construct a linkage map showing the percentage recombination values between these mutant sites.

Answer 6.9

Remember that on *E. coli* K only *rII*$^+$ recombinants form plaques, so that the best estimate of the total number of recombinants is twice the number of plaques on *E. coli* K. Note also that no dilution factor is involved in these crosses.

% recombination values:

$$rIIa \times rIIb \quad = \frac{2 \times 9}{390} \times 100 = 4.6\%$$

$$rIIb \times rIIc \quad = \frac{2 \times 8}{460} \times 100 = 3.4\%$$

$$rIIa \times rIIc \quad = \frac{2 \times 13}{306} \times 100 = 8.5\%$$

Ordering the map so that the map distances are approximately additive, the only possible order is

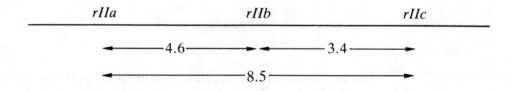

Question 6.10

rI mutants of phage T4 form large clear (rapid lysis) plaques on all strains of *E. coli*, whereas *tu* (*tu*rbid) mutants form small turbid plaques. In a cross between an *rI* and a *tu* mutant the following plaques were scored when a lysate from a doubly infected host was plated on *E. coli* K:

large clear	1482
small turbid	1515
large turbid	389
small clear	364 (total 3750)

What is the percentage recombination between *rI* and *tu*?

Answer 6.10

The large clear and the small turbid plaques are formed by the non-recombinant parental phages (*rI tu*+ and *r*+ *tu*, respectively), while the large turbid and small clear plaques are produced by the recombinants (*r tu* and *r*+ *tu*+, respectively). Hence, the percentage recombination between *rI* and *tu* is

$$\frac{389 + 364}{3750} \times 100 = 20\%$$

NOTE

In this cross, unlike crosses between *rII* mutants, it is not possible to select a particular class of phage recombinant; both classes of parental and both classes of recombinant phage progeny are detectable, and percentage recombination frequencies are calculated by use of the standard formula.

6.6 Supplementary Questions

6.1 Why are the recombinants from Hfr × F⁻ crosses nearly always F⁻?

6.2 Explain why different Hfr strains have different origins and directions of transfer.

6.3 Show how you would isolate an F′$^{gal+}$ strain from an Hfr strain of *E. coli*.

6.4 In transductions it is always necessary to plate separately onto the selective medium a sample of the recipient bacteria and a sample of the phage suspension grown on the donor. Why?

6.5 You have isolated a new F′ factor. How could you most easily show that part of the original F factor DNA has been deleted and replaced by a segment of 'foreign' (i.e. bacterial) DNA?

6.6 What may happen to a fragment of donor DNA introduced into a recipient cell by transduction?

6.7 In a transformation experiment a^-b^- recipient cells were transformed with either DNA from an a^+b^+ donor *or* a mixture of DNA extracted from a^+b^- and a^-b^+ cells. The results were as follows:

	Transformants per ml		
	a^+b^+	a^+b^-	a^-b^+
a^+b^+ donor DNA	30	1500	2000
a^+b^- and a^-b^+ donor DNA	25	1400	1800

What can you infer about the linkage relationships between *a* and *b*?

6.8 Which method of gene transfer would you use to determine (**a**) the location on the *E. coli* chromosome of a newly discovered locus and (**b**) the position within a gene of a newly isolated mutation? State your reasons.

6.9 When *E. coli* strain K is simultaneously infected with two particular *rII* mutants of phage T4, the infected cells lyse and release normal numbers of T4 progeny. Are these progeny the result of recombination or complementation? How would you distinguish between these alternatives?

6.10 In *E. coli* two loci, *A* and *B*, are separated by about 5% of the genome; two other markers, *C* and *D*, are separated by only 1% of the genome. Which pairs of markers could be (**a**) co-transformed and (**b**) co-transduced?

7 Gene–protein relationships

7.1 Proteins

Proteins play a great diversity of essential roles in all living cells; in the comparatively simple bacterium *E. coli* there are about 2000 different types of protein, while in humans the number is estimated at 100 000. Many proteins play largely structural roles; thus, collagen, a fibrous protein, is an important component of bones, tendons and ligaments, myosin is the main muscle protein, and the protein coat of a virus encapsulates and protects the viral nucleic acid. Other proteins, such as the myoglobins and haemoglobins, are involved in oxygen transport and storage, while hormones regulate various types of chemical activity in the cell. Many proteins are enzymes, such as DNA polymerase and RNA polymerase, biological catalysts which promote or speed up specific biochemical reactions in the cell without becoming changed themselves, so that a single molecule of an enzyme can catalyse the synthesis or breakdown of many hundreds of substrate molecules. Each enzyme has an active site which binds the specific substrate molecules, forming an enzyme–substrate complex; this brings closer together the reactive groups of the substrate molecules and promotes the formation of covalent bonds between them. After the covalent bonds have formed, the products are released and the enzyme molecule recycles. Each enzyme is highly specific, and usually, catalyses one specific biochemical reaction.

7.2 Polypeptides

All proteins are complex macromolecules assembled from basic structural units called **amino acids**. Each amino acid (Figure 7.1) has an amino group (—NH$_2$), a carboxyl group (—COOH) and a specific side-chain (the radical or R group) attached to the same carbon atom. There are 20 commonly occurring L-amino acids and all are found incorporated into protein.

These amino acids are enzymatically joined together by a covalent bond forming between the carboxyl group of one amino acid and the amino group of another, forming a **peptide bond** and releasing a molecule of water (Figure 7.2). Very many amino acids can be joined in this way, forming a long linear molecule, a **polypeptide chain** or **polypeptide**; an 'average' polypeptide is about 250–300

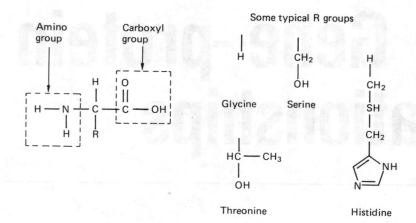

Figure 7.1 *The general structure of amino acids. Amino acids differ only by the radical (R) group attached to the central carbon atom*

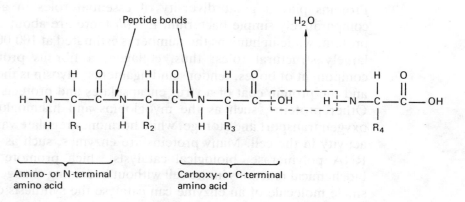

Figure 7.2 *A polypeptide chain. A fourth amino acid is about to be added on to the C-terminal residue by the formation of a peptide bond; this releases a molecule of water. Note that there is a regularly repeating backbone chain and variable side-chains (R groups). Chain extension only occurs at the carboxy terminal and amino acid sequences are always read in the amino terminal to carboxy terminal direction*

residues long ('residue' is the term used to refer to an amino acid when it is incorporated into a polypeptide).

These polypeptides are the precursors of proteins. Each different polypeptide is the end product of the activity of a specific structural gene and has a *specific amino acid sequence*, its **primary structure**, and this, in turn, is determined by the sequence of nucleotides along the structural gene encoding it — as we shall see in Chapters 8 and 9, the sequence of nucleotides in the DNA (the genetic code) is transcribed and translated into the specific sequence of amino acids in a polypeptide. Just how the polypeptide chains fold and aggregate together to form an active protein is *determined solely by their primary structure*. This process is of critical importance, since the activity of a protein is totally dependent on the spatial arrangement of the amino acids within the three-dimensional structure and the substitution of just one amino acid for another may completely alter the structure and activity of a protein.

Polypeptides are commonly folded into a regularly repeating configuration, the **secondary structure**. The most common secondary structures are (1) the α **helix**, a spiral chain of amino acids held together by weak hydrogen bonds between the CO and NH groups of adjacent amino acids, and (2) the β structure, where two or more polypeptide strands are aligned side by side and held together by H bonds between the CO and NH groups on *different* strands; β structures may be formed either by folding back and forth a single polypeptide chain or by the interactions between many separate polypeptide chains to form a β sheet.

Many of the most important proteins have **globular** shapes (these include the enzymes, haemoglobins and regulatory proteins) and their secondary structure typically consists of α helixes interspersed with β structures and random folds. These structures are now folded back and forth to form compact globular structures; this folding is known as the **tertiary structure**. Several types of interaction stabilise the tertiary structure, including the formation of disulphide bridges between physically adjacent cysteine residues (Figure 7.3) and weak hydrogen bonding. This structure is very important, as it is responsible for the specific activities of a protein.

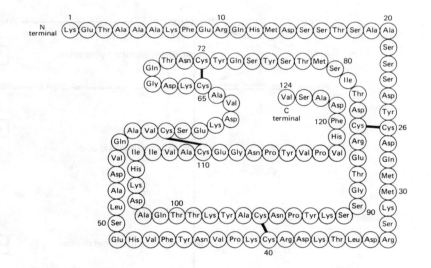

Figure 7.3 *Bovine pancreatic ribonuclease. This consists of a single polypeptide, 124 residues long, folded into a secondary structure held together by four disulphide bridges (black bars)*

Some proteins consist of two or more polypeptide chains, or **sub-units**, aggregated together in a highly specific manner. These sub-units may be either similar or different; thus, the *E. coli* enzyme β-galactosidase is a tetramer of four identical polypeptides each 1024 residues long, while human adult haemoglobin consists of two identical α chains, each 141 residues long, and two identical β chains, each 146 residues long. These aggregates are **quaternary structures**.

7.3 Inborn Errors of Metabolism

The first relationship between a gene and a specific protein was suggested by Sir Archibald Garrod in his book *Inborn Errors of Metabolism*, published in 1909. As a result of studies made by himself and by William Bateson (1902), the human disease **alkaptonuria** was known to be caused by homozygosity for a recessive Mendelian gene. This rare and comparatively mild disorder (it affects about one person in every million) results in a mild arthritic condition in later life and in the urine turning black on exposure to air, owing to the accumulation of homogentisic acid and its subsequent oxidation to a melanin-like substance. Garrod correctly concluded that homogentisic acid is one of the breakdown products of the amino acids phenylalanine and tyrosine and, that in normal individuals, this is broken down before being excreted in the urine as fumaric and acetoacetic acids (Figure 7.4). Garrod conceived that this reaction was the work of a special enzyme (then called a ferment) which was lacking in congenital alkaptonuriacs. Thus, alkaptonuria, and other inborn errors of metabolism, could be interpreted as a block at a particular point in normal intermediary metabolism due to the absence of a genetically determined enzyme — a particular step in a biochemical pathway could not take place. This enzyme was only identified in 1958 and proven to be absent in alkaptonuriac patients.

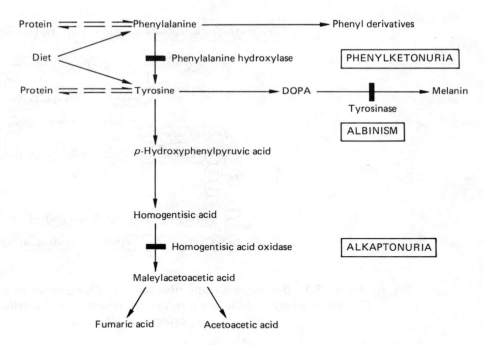

Figure 7.4 *The metabolic pathway for the degradation of phenylalanine and tyrosine in man. The bars indicate the positions of the genetic blocks in the pathway*

Several other diseases are the result of homozygosity for different genes leading to the absence of other enzymes in the same metabolic pathway (Figure 7.4). Two of the most important are:

(1) **Phenylketonuria (PKU)** This disease affects about 1 in 10 000 newborn infants and, if untreated, leads to severe mental retardation. The missing enzyme is phenylalanine hydroxylase and this leads to the accumulation of phenyl-alanine; this is converted to other phenyl derivatives (including phenylpyruvate) which enter the cerebrospinal fluid and cause irreversible brain damage. If detected at birth, the disease can be treated by feeding the infants on a low-phenylalanine diet (but note that some phenylalanine and tyrosine must be present in the diet, since these amino acids cannot be made by humans but are required for incorporation into protein). This is why the urine of all newborn infants is routinely screened for increased levels of phenylpyruvic acid (the PKU test).

(2) **Recessive albinism** This condition, also studied by Garrod, affects about 1 in every 20 000 individuals and results in the absence of pigmentation in the skin and eyes. Affected individuals lack a tyrosinase which converts DOPA (3:4-dihydroxyphenylalanine) into the pigment melanin.

7.4 The One-gene, One-enzyme Hypothesis

During the 1930s the work of Garrod was confirmed and extended by George Beadle and Boris Ephrussi working on eye-colour mutants in *Drosophila*. Using a series of mutants, they were able to show that the synthesis of the brown pigment present in normal eyes proceeds along a biochemical pathway and that each step in the pathway is under the control of a single gene; blocks at different steps in the pathway (caused by homozygosity for one of the several recessive eye-colour alleles) led to the accumulation of different intermediates and to the production of an altered eye colour (Question 7.1).

These experiments led George Beadle, now working in collaboration with Edward Tatum, to study the hypothesised relationships between genes and enzymes in more detail, using the Ascomycete *Neurospora crassa*. They knew that *N. crassa* was able to grow on a simple chemically defined **minimal medium;** this contains certain inorganic salts (including a nitrogen source), a sugar as a carbon source and the vitamin biotin, from which the wild type or **prototrophic** strain could synthesise all the other vitamins, amino acids and nucleotide bases required for growth and reproduction. They reasoned that it should be possible to isolate **biochemical mutants** unable to carry out a particular step in a particular pathway; such mutants, or **auxotrophs**, would only grow on minimal medium if it were supplemented with either the end product of that particular pathway or an intermediate occurring subsequent to the block. For example, we now know that the amino acid arginine is synthesised along the pathway

$$N\text{-acetylornithine} \rightarrow \text{ornithine} \rightarrow \text{citrulline} \rightarrow \text{arginine}$$
$$\qquad\qquad\quad 1 \qquad\qquad\quad 2 \qquad\qquad 3$$

Thus, a mutation in a gene controlling the production of the enzyme responsible for step 1 would produce an auxotroph able to grow on minimal medium supplemented with ornithine, citrulline or arginine, while a different mutant blocked at step 3 could only grow if arginine was provided in the medium.

Beadle and Tatum isolated many such auxotrophs and they proposed that each gene acted by storing the genetic information required for the production of a particular enzyme — this was the first statement of the **one-gene, one-enzyme** hypothesis, now more correctly referred to as the **one gene, one polypeptide** hypothesis (Question 7.2).

These discoveries opened up the new field of biochemical genetics and it was by this type of combined biochemical and genetic analysis that many of the cellular biosynthetic pathways were first elucidated (Question 7.7)

Very soon after the isolation of biochemical mutants in *Neurospora* Edward Tatum isolated similar mutants in the bacterium *Escherichia coli*, thus opening up the field of bacterial genetics.

7.5 Genes Function by Determining the Amino Acid Sequences of Proteins

The foregoing experiments clearly demonstrated a close correlation between the presence (or absence) of a particular gene and the presence (or absence) of a particular enzyme, but they did not show that genes acted by determining the amino acid sequences of polypeptides.

This was first shown by Vernon Ingram in 1956, when he found that each mutation occurring in the gene encoding the β-polypeptide of human haemoglobin A caused the substitution of one amino acid for another in the β-polypeptide (Question 7.3). An even more precise relationship comes from the work of Charles Yanofsky on one of the proteins making up the complex *E. coli* enzyme, tryptophan synthetase. By 1957 he had shown that there was a direct correspondence between the sequence of mutations within the gene encoding the A polypeptide (the *trpA* gene) and the sequence of amino acids within the A polypeptide itself — this was the first demonstration that a gene and the protein it encodes are **colinear** (Question 7.4).

7.6 Complementation

Many genetic studies commence with the isolation of a set of mutants, such as a series of eye-colour mutants in *Drosophila*, or of a particular auxotrophic phenotype in *Neurospora* or *E. coli*, and although recombination analysis will establish the spatial relationships between the different mutations, it tells us nothing about their functional relationships. If any two particular mutations turn out to be unlinked, or only loosely linked, then clearly two different or **non-allelic** genes are involved, but if the two mutations cause similar phenotypes and are closely linked, then they may or may not be allelic. This question can usually be resolved by testing whether or not the mutants **complement** each other.

(a)   The *cis–trans* Test in Diploid Organisms

In diploids this is achieved by using the *cis–trans* test. If *m1* and *m2* are two independently isolated recessive mutations affecting the same phenotypic char-

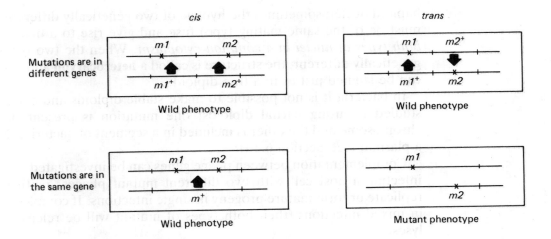

Figure 7.5 *The complementation test. If two mutations are in different structural genes, the* trans *arrangement produces a wild phenotype, but if they are in the same gene, the* trans *arrangement results in a mutant phenotype. The arrows indicate the presence of the wild type gene product*

acter, complementation is detected by comparing the phenotypes of the *cis* and *trans* double heterozygotes. In the *cis* arrangement (Figure 7.5) both mutations are on one chromosome and the other chromosome is wild type; in the *trans* arrangement one mutation is present on each chromosome.

If the two mutations occurred in different genes (i.e. are non-allelic), then both the *cis* and *trans* arrangements will produce a wild phenotype, but if they occurred in the same gene (i.e. allelic), the *trans* arrangement produces a mutant phenotype, since neither chromosome carries a wild type gene, and only the *cis* arrangement has a wild phenotype. When the *trans* arrangement produces a wild phenotype, the mutants are said to **complement** each other, the defect in each mutant gene being made good, or complemented, by the corresponding wild type genes on the homologous chromosomes. This type of complementation occurs between genes and is known as **intergenic complementation**.

(b) Complementation in Micro-organisms

The complementation test requires two copies of the gene or genes being investigated to be present in the same cell. Since all higher eucaryotes are normally diploid, it is only necessary to make the correct crosses, but some lower eucaryotes and all procaryotes are normally haploid, and it is more difficult to produce a diploid or partially diploid state.

In the yeast *Saccharomyces cerevisiae*, a unicellular ascomycete, testing is simple because the vegetative cells can exist in both haploid and diploid phases, and diploids can be constructed by allowing two haploid strains of opposite mating type to undergo cellular and nuclear fusion (Figure 5.3). In other ascomycetes, such as *Neurospora*, a true diploid cell is only formed immediately prior to meiosis and sporulation, and complementation is studied by forming heterocaryons. *N. crassa* has a filamentous mycelium composed of branched hyphae divided by cross-walls into many compartments each containing many haploid nuclei; sometimes the hyphae of two genetically different strains (which must be of the same mating type) fuse and give rise to a mycelium *containing both types of nuclei in a common cytoplasm*. When the two types of nuclei are genetically different, the structure is called a **heterocaryon** and complementation can be studied just as in a true diploid.

In bacteria it is not possible to make stable diploids and complementation is studied by using partial diploids; one mutation is present on the bacterial chromosome and the other is included in a segment of bacterial DNA carried on a plasmid (see Section 6.3d).

Complementation between phage genes can be investigated by simultaneously infecting a host cell with two different mutant phages, neither of which can replicate or form mature progeny in single infections. If complementation occurs in mixed infections, then both types of mutant will be released when the cell lyses.

(c) Limitations of Complementation Testing

Except in certain rare instances where either recombination has occurred within a gene or particular pairs of mutants within the same gene complement each other (**intragenic complementation**), a positive result in a complementation test shows (1) that two non-allelic genes are involved and (2) that each acts by producing a diffusible gene product.

However, a negative result does not necessarily indicate that the two mutations are allelic; the most likely exception is when one of the mutants (at least) involves a genetic control sequence (Question 10.1).

7.7   Feeding Tests

Complementation is a simple genetic test demonstrating that the same phenotype can be caused by mutations in two or more non-allelic genes. In certain instances, particularly in micro-organisms, two further tests can be used to establish non-allelism between two phenotypically similar mutants.

Consider, for example, the biochemical pathway for the synthesis of arginine in *N. crassa* (Section 7.4). Arginine-requiring mutants can result from mutations in any one of at least three different genes — those controlling steps 1, 2 and 3 in the biosynthetic pathway. If mutation has occurred in a gene affecting step 2, then ornithine will accumulate in the mutant strain and it will only be able to grow on minimal medium when either citrulline or arginine is provided; similarly, a mutant blocked in step 3 will accumulate citrulline and can only grow if arginine is provided. Thus, the two mutants can be distinguished by the intermediates upon which they will grow.

The second test, called a **feeding test**, does not require a prior knowledge of the intermediates in the biochemical pathway. Consider a feeding test where we plate side-by-side on minimal medium mutants blocked in steps 2 and 3. The mutant blocked in step 3 will accumulate citrulline and this will diffuse into the adjacent medium, where it can be utilised by the mutant blocked in step 2; there will not be a lot of growth but there will be enough to be easily detectable (Figure 7.6).

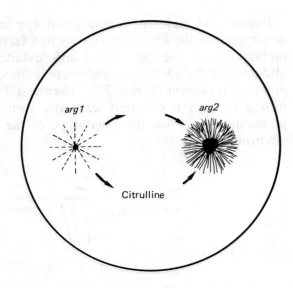

Figure 7.6 *Feeding. The slight growth of the* arg1 *mutant allows some citrulline to accumulate and to diffuse into the medium; this allows growth of the* arg2 *mutant, which will grow when either citrulline or arginine is provided*

The important point to note is that the block in the mutant that is fed is always before the block in the mutant that is feeding.

It was by this type of combined genetic and biochemical analysis that many of the biochemical pathways in living cells were first elucidated.

7.8 Questions and Answers

Question 7.1

Describe how studies with eye-colour mutants of *Drosophila* supported the concept of a relationship between a gene and a particular protein.

Answer 7.1

In *Drosophila* many of the structures of the adult fly develop from a variety of imaginal discs present in the larva; these are undifferentiated groups of cells and, upon metamorphosis, each disc develops into a specific adult structure.

In adult flies the eye colour is a dull red, due to a combination of unrelated red and brown pigments, and in the 1930s George Beadle and Boris Ephrussi set out to study the biochemical pathway leading to the production of the brown pigment. They used a number of recessive eye-colour mutants and two of these, *v*, vermilion, and *cn*, cinnabar, were the most important. Both *v/v* and *cn/cn* homozygotes had nearly identical bright red eyes due to the absence of the brown pigment; however, the two mutants had different defects in brown pigment production.

Beadle and Ephrussi transplanted eye imaginal discs from a larva of one genotype into the abdominal cavity of a larva with a different genotype; during metamorphosis the disc would differentiate into an eye-like structure in the abdomen of the adult fly, from where it could be dissected and examined for the pigment(s) present (Figure 7.7). They asked whether a mutant disc transplanted into a host of a different genotype would show mutant or wild type eye pigmentation — was the phenotype of the implanted disc altered by its new environment?

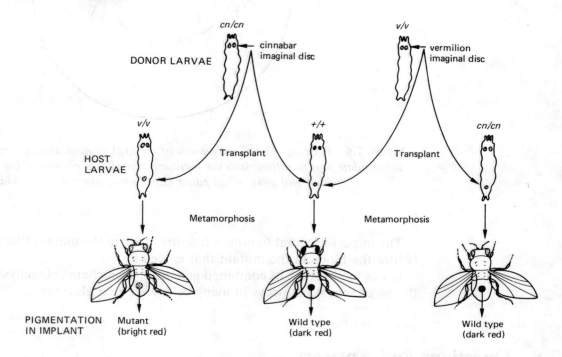

Figure 7.7 *The Beadle and Ephrussi imaginal disc transplantation experiments in* Drosophila

They found:

(1) *v/v* and *cn/cn* discs implanted into +/+ hosts both developed wild type eye colour.
(2) *v/v* discs implanted into *cn/cn* hosts developed wild type eye colour.
(3) *cn/cn* discs implanted into *v/v* hosts developed mutant eye colour.

They reasoned that the brown pigment is synthesised along a biochemical pathway involving at least two separate steps:

$$A \xrightarrow{\;1\;} B \xrightarrow{\;2\;} \text{brown pigment}$$

and that the *vermilion* locus controls one of these steps and the *cinnabar* locus the other, probably by being responsible for the production of the enzymes catalysing these reactions. Furthermore, since *v/v* transplants form brown pigment in *cn/cn* hosts, while *cn/cn* transplants produce only red pigment in *v/v* hosts, they reasoned that the *cn/cn* host must produce a diffusible compound which can be utilised by the *v/v* transplant and converted into a brown pigment, so that the *vermilion* locus must control an earlier step in the pathway than the *cinnabar* locus. Thus, the diffusible intermediate accumulating in the *cn/cn* host could be converted into a brown pigment by the activity of the *cn*+ gene product in the transplant.

NOTES

1 Other mutants studied by Beadle and Ephrussi did not behave in this way and the pigment produced in the transplant was determined exclusively by the genotype of the donor larva. The reason for this is not clear, but it is possible that the intermediates accumulating in the host were not diffusible and so never reached the transplanted disc.

2 These experiments were of great importance, since they not only confirmed the ideas put forward by Garrod in 1909, but also encouraged Beadle, in collaboration with Edmund Tatum, to commence investigations using nutritional mutants of the fungus *Neurospora crassa* — studies that led to the formulation of the one gene, one enzyme hypothesis.

3 Subsequent work established the biochemical pathway shown in Figure 7.8. The position of the blocks in *cn* and *v* mutants has been confirmed by feeding experiments. For example, *v/v* larvae (but NOT *cn/cn* larvae) develop wild type eye pigmentation if they are fed with or injected with either formylkynurenine or kynurenine; thus, in *v/v* larvae the biochemical block is prior to these intermediates.

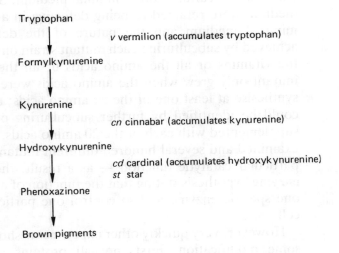

Figure 7.8 *The production of brown eye pigment in* Drosophila. *The position of the blocks caused by some of the eye-colour mutations are shown at the right*

Question 7.2

Write a short essay on the 'one gene, one enzyme' hypothesis of Beadle and Tatum.

Answer 7.2

By the early 1940s it was known that complex biochemical compounds, such as nucleotides, amino acids, vitamins and the other molecules required for growth and reproduction, were synthesised by a series of small steps, each step catalysed by its own enzyme and the product of each step being passed on to the next enzyme so that the following step could take place:

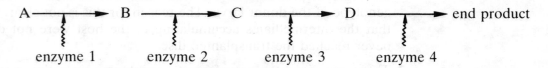

In 1941 George Beadle and Edward Tatum argued that, if the enzymes involved in such a **biochemical pathway** were encoded by genes, it should be possible to isolate mutant strains lacking one or other of these enzymes and these would be unable to carry out one or other of the reactions along the pathway; such mutants would only grow if the end product of the pathway were provided in the medium. They hoped to see how genes worked by isolating mutations in them and then ascertaining the nature of the biochemical defect.

They used *Neurospora crassa*, an ascomycete able to grow on a simple minimal medium containing only inorganic salts, a carbon source (a sugar) and the vitamin biotin.

They started by irradiating the asexual spores with either X-rays or ultraviolet light so as to speed up the process of mutation. The irradiated spores were then grown on a fully nutrient medium and, after confirming the purity of each strain, each was subcultured onto minimal medium. Strains failing to grow on minimal medium were retained as being defective in a function essential for growth on minimal medium and the nature of the defect was ascertained. This they achieved by subculturing each mutant strain onto minimal medium containing all the vitamins or all the amino acids or all the nucleotides. If, for example, a mutant only grew when the amino acids were provided, then it was unable to synthesise at least one of the 20 amino acids; the particular amino acid missing could be identified by further subculturing onto minimal medium separately supplemented with each of the 20 amino acids. Over 68 000 treated spores were examined and several hundred different mutants isolated, each one lacking one particular catalytic function — as a result, they proposed their **one gene, one enzyme** hypothesis stating that the function of a gene is to direct the formation of one specific enzyme and so control one particular biochemical reaction in the cell.

However, very quickly other experiments showed that the hypothesis required some modification. First, not all proteins are enzymes (for example, the haemoglobins and collagen), so that it was more accurate to say **one gene, one protein**. But even this is not correct, as some proteins are complexes of two or more different polypeptides, each encoded by a separate gene; a more correct restatement of the hypothesis is **one gene, one polypeptide chain**. Finally, we must remember that not all structural genes (and here we are only discussing structural genes) encode polypeptides; some genes encode the species of RNA involved in the process of translation (transfer RNA and ribosomal RNA).

Question 7.3

How did studies on human haemoglobins first demonstrate that a gene acts by determining the amino acid sequence of a protein?

Answer 7.3

In normal adult humans the red blood cells contain haemoglobin A (HbA), a protein made up from two alpha (α) and two beta (β) polypeptides and essential for transporting oxygen around the body. In one particular inherited disease, sickle-cell anaemia, the red blood cells contain an abnormal haemoglobin (HbS) and are very inefficient carriers of oxygen, and about 50% become sickle-shaped whenever oxygen tension is low. These individuals suffer from a progressive haemolytic anaemia frequently resulting in early death.

In 1949 James Neel and E. A. Beet independently demonstrated that the disease is inherited and that affected individuals are homozygous for a recessive mutant allele, Hb^S. Heterozygous individuals (Hb^A/Hb^S) are carriers of the disease and can transmit the harmful gene; they exhibit the **sickle-cell trait** — the red blood cells may show moderate sickling (about 1%) but the individuals do not suffer from the anaemia.

In the same year Linus Pauling showed that the haemoglobins from normal (HbA) and affected (HbS) individuals migrated at different rates in an electric field and that both types of haemoglobin were present in individuals with the trait, and he considered sickle-cell anaemia to be a gene-controlled molecular disease.

The molecular nature of this difference was shown by Vernon Ingram in 1957. He demonstrated that the only difference between HbA and HbS was the sixth residue in the beta chain; in HbA this residue is glutamic acid, whereas in HbS it is valine.

The significance of this work is that it clearly demonstrated that mutation in the HbA genes can dramatically affect the phenotype by causing a single amino acid substitution.

Subsequently, many different HbA variants have been found and all differ from the wild type haemoglobin by a single amino acid substitution in either the alpha chain or beta chain.

NOTE

Another mutant allele causes haemoglobin C (HbC) disease, a comparatively mild form of anaemia; its particular interest is that the same glutamic acid residue in the beta chain has been replaced, but by a lysine. Other abnormal haemoglobins do not cause any observable clinical symptoms; clearly, the ability of a protein to have normal biological activity depends on the polypeptides having specific amino acid residues at some positions, while at other positions the particular residue present is not important.

Question 7.4

What is meant by the colinearity of a gene and the protein it encodes? Describe one contemporary experiment demonstrating this relationship.

Answer 7.4

Colinearity refers to the correlation between the sequence of nucleotides within a gene and the sequence of amino acids in the corresponding polypeptide.

In the 1960s Charles Yanofsky and his colleagues set out to establish this relationship by showing that there is a direct correlation between the order of mutant sites within a particular gene and the correspondingly altered amino acid sequence in the polypeptide encoded by that gene.

They chose to study one of the genes encoding the complex enzyme tryptophan synthetase in *E. coli*. This enzyme catalyses the conversion of indole-3-glycerol phosphate (IGP) and serine into tryptophan, the final step in the pathway for tryptophan biosynthesis, and consists of four easily separable sub-units, two α-chains and two β-chains; these polypeptides are encoded by the *trpA* and *trpB* genes, respectively, and are 267 and 397 amino acids long. Their experimental procedure was as follows:

(1) They isolated a set of 16 mutants within the *trpA* gene. These mutants (a) could only grow if tryptophan was provided in the growth medium and (b) contained an altered α-polypeptide, still able to complex with the β-polypeptide but unable to convert IGP and serine into tryptophan. Nevertheless, these enzyme complexes could, *in vitro*, convert indole and serine into tryptophan; this method of selecting mutants ensured that a complete, albeit altered, α-polypeptide was present.

(2) The order of the mutant sites within the *trpA* gene was established by recombinational analysis.

(3) The altered α-polypeptides were isolated from each mutant and their amino acid sequences determined.

(4) The sequence of mutational sites within *trpA* was compared with the sequence of the altered amino acids in the α-polypeptide.

They found that there was a complete correspondence between the sequence of the mutational sites within *trpA* and the sequence of amino acids in the corresponding protein (Figure 7.9).

Question 7.5

Several loci on the X chromosome of *Drosophila melanogaster* control the colour of the eye. Two eye-colour mutants, raspberry (*ras*) and prune (*pn*), result in

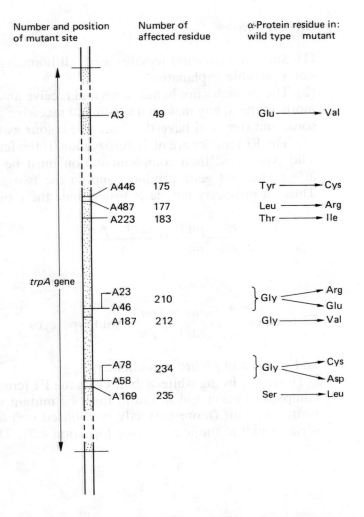

Number and position of mutant site	Number of affected residue	α-Protein residue in: wild type	mutant
A3	49	Glu ⟶	Val
A446	175	Tyr ⟶	Cys
A487	177	Leu ⟶	Arg
A223	183	Thr ⟶	Ile
A23 / A46	210	Gly ⟶	Arg / Glu
A187	212	Gly ⟶	Val
A78 / A58	234	Gly ⟶	Cys / Asp
A169	235	Ser ⟶	Leu

trpA gene

Figure 7.9 *Colinearity of gene and protein. There is a direct correspondence between the order of the mutant sites within the* trpA *gene and the positions of the amino acid substitutions within the α-protein*

dark ruby eyes in homozygous females and hemizygous males; two further X-linked mutations produce white (*w*) and buff (*bf*) eyes. The table below shows the eye colour of the female progeny in crosses between females and males with the eye colours indicated (note that the male progeny always had the same eye colour as their mother). Explain these results and deduce the minimum number of genes involved in the production of the eye pigments.

		EYE COLOUR OF FATHER			
		raspberry	prune	white	buff
EYE COLOUR OF MOTHER	raspberry	raspberry	wild type	wild type	wild type
	prune	wild type	prune	wild type	wild type
	white	wild type	wild type	white	buff
	buff	wild type	wild type	buff	buff

eye colour of female progeny

123

Answer 7.5

(1) Since the parental females were all homozygous, meiotic recombination is not a possible explanation.

(2) The F1 males are hemizygous and receive an X chromosome only from their mothers; thus, any maternal sex-linked recessive genes will be expressed in these sons, and they will have the same eye colour as their mothers.

(3) The F1 females are all heterozygous. If the females in a particular cross have wild type eyes, then complementation must be occurring, the two mutations affect different gene products and so the two genes involved are non-allelic. Thus, in raspberry female × prune male the cross must have been

$$\frac{ras \quad pn^+}{ras \quad pn^+} \times \frac{ras^+ \quad pn}{Y}$$

$$\downarrow$$

F1 females $\qquad \dfrac{ras \quad pn^+}{ras^+ \quad pn}$ wild type eyes

and so *ras* and *pn* are non-allelic.

However, in the white × buff cross the F1 females had buff eyes. There is no complementation and *w* and *bf* must be mutant within the same gene. In fact, white and buff (more correctly symbolised w^{bf}) are part of the multiple allelic series found at the *w* locus (see Question 4.7). Thus,

$$\frac{w}{w} \times \frac{w^{bf}}{Y}$$

$$\downarrow$$

F1 females $\qquad \dfrac{w}{w^{bf}}$ buff eyes

Similar reasoning establishes that prune and raspberry are not allelic with the white gene. Hence, at least three loci, *ras, pn* and *w*, all located on the X chromosome, are involved in the production of eye pigments.

NOTE

Where two mutations occur at different sites within the same gene, it is possible for recombination to occur between them and reconstruct a wild type gene, either at meiosis or at a mitosis during development:

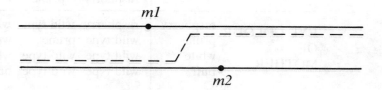

In this example this could only happen at a mitosis during the development of individual female flies. Since it is a rare type of event, only occasional female flies would show any wild type pigmentation, and even then, depending on when the recombination event occurred during development, only one eye or perhaps just a sector within an eye would be normally pigmented. On the other hand, with complementation **every** heterozygous female will have fully pigmented eyes.

Question 7.6

You have isolated six tryptophan-requiring mutants of E. coli and, by recombinational analysis, have shown that all the mutant sites are closely linked. You then carried out a complementation analysis by making partial diploids between each pairwise combination of mutants and plating these on minimal medium; the results of these tests were as follows:

mutant	trp-1	trp-2	trp-3	trp-4	trp-5	trp-6
trp-1	0					
trp-2	+	0				
trp-3	+	+	0			
trp-4	+	+	+	0		
trp-5	+	+	0	+	0	
trp-6	+	+	+	+	+	0

(a + indicates growth on minimal medium, a 0 the absence of growth)

What is the minimum number of genes involved in the biosynthesis of tryptophan?

Answer 7.6

(1) No mutant complements itself (as expected).
(2) Among the dissimilar pairwise combinations, only mutants trp-3/trp-5 fail to complement. Therefore, it is probable that the trp-3 and trp-5 mutations occurred within the same gene.
(3) All other pairwise combinations complement, so that the minimum number of genes involved in tryptophan biosynthesis is FIVE.

Question 7.7

Continuing the experiments referred to in Question 7.6, you (**a**) tested to see which mutants grew on anthranilic acid (AA), phosphoribosyl anthranilate (PRA), indoleglycerol phosphate (IGP) and tryptophan (Trp) and (**b**) determined which pairs of mutants resulted in cross-feeding. The results of these tests were:-

(a) Growth on

	AA	PRA	IGP	Trp
trp-1	−	−	+	+
trp-2	+	+	+	+
trp-4	−	−	−	+
trp-5	−	+	+	+
trp-6	−	−	−	+

(b) Feeding tests

trp-1 was fed by trp-4 and trp-6
trp-2 was fed by trp-1, trp-4, trp-5 and trp-6
trp-4 was not fed by any other mutant
trp-5 was fed by trp-1, trp-4 and trp-6
trp-6 was not fed by any other mutant

In these tests *trp-3* behaved identically with *trp-5*.
What is the sequence of reactions in the biosynthetic pathway for tryptophan?
What can you say about the genes and enzymes involved in these reactions?

Answer 7.7

(a) A mutant blocked at a particular step in a biochemical pathway is ONLY fed by another mutant blocked at a later stage in the same pathway.

Since *trp-2* is fed by all the other mutants (*trp-3* and *trp-5* can be considered together), this mutation must have occurred in a gene controlling the first step in the pathway. Since *trp-5* is fed by *trp-1*, *trp-4* and *trp-6*, while *trp-1* is fed only by *trp-4* and *trp-6*, the second step is controlled by the *trp-3* gene, the third step by the *trp-1* gene and the final step by the *trp-4* and *trp-6* genes.

Thus, the sequence of the gene controlled reactions is

$$\longrightarrow trp\text{-}2 \longrightarrow trp\text{-}5 \longrightarrow trp\text{-}1 \longrightarrow \begin{array}{c} trp\text{-}4 \\ trp\text{-}6 \end{array} \longrightarrow$$

(b) Likewise, a mutant blocked in an early step in the pathway can only grow when intermediates occurring later in the pathway are provided. Since only one mutant grows on AA, two grow on PRA, three grow on IGP and all grow on Trp, the biochemical pathway and the position of the blocks caused by each mutant is

$$\xrightarrow{\quad \|\quad} AA \xrightarrow{\quad \|\quad} PRA \xrightarrow{\quad \|\quad} IGP \xrightarrow{\quad \|\quad} Trp$$

trp-2   trp-5   trp-1   trp-4
     trp-6

This experiment shows that *at least one gene* is involved in each of the first three reactions and *at least two genes* in the conversion of IGP to Trp. Only the study of a larger series of mutants would establish whether there are any other steps in the pathway for tryptophan biosynthesis and how many different polypeptide sub-units make up each of the enzymes involved.

NOTE

The enzyme converting IGP to tryptophan is known as tryptophan synthetase and it is a complex of two different protein sub-units encoded by the *trpA* and *trpB* genes. The *trpA* protein was studied by Yanofsky in his experiments on the colinearity of a gene and the polypeptide it specifies — see Question 7.4 and Section 7.5.

Question 7.8

Distinguish between intergenic and intragenic complementation.

Answer 7.8

Complementation is the ability of two recessive mutants to make good each other's defect when they are present in the same cell but on different chromosomes (i.e. in the *trans* arrangement). Complementation is intergenic when the two mutations are in different genes and intragenic when they are in the same gene; thus, intragenic complementation is an exception to the rule that complementation normally occurs only between non-allelic mutants. An important functional distinction is that with intergenic complementation *all* the mutants with mutations in one gene will complement *all* the mutants with mutations in a different gene, whereas intragenic complementation only occurs between *particular* pairs of allelic mutants and is only detectable within certain genes (see next question).

Question 7.9

How can intragenic complementation be explained in terms of protein–protein interaction?

Answer 7.9

Many proteins are aggregates of two or more *identical* polypeptide chains (homopolymers), and if mutation causes an amino acid substitution in the protein sub-unit, it may alter the three-dimensional configuration of the sub-unit so that it can no longer aggregate. Different amino acid substitutions will produce mutant subunits with different three-dimensional configurations, none of which can aggregate with itself to form an active enzyme. It is thought possible that certain pairs of mutations within the gene encoding the sub-unit give rise to sub-units with complementary configurations which *are* able to aggregate in dissimilar pairs to form an active or partially active enzyme. Thus, a diploid cell or organism heterozygous for this particular pair of mutations will form some active enzyme. The enzyme produced may have reduced catalytic activity and, since only 50% of the pairs of sub-units can complement, will be present in less than normal amounts (Figure 7.10).

NOTE

1 Intragenic complementation can only be detected between alleles of a gene encoding a polypeptide which then aggregates to form an active homopolymer.

2 Most pairs of mutants within such a gene will not complement.

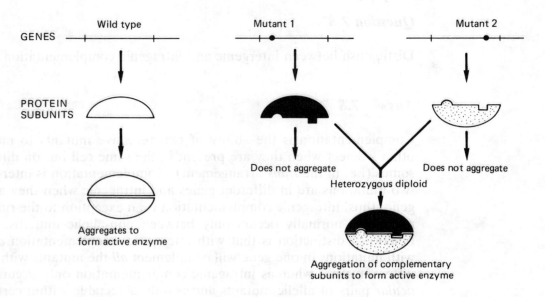

GENES

Wild type   Mutant 1   Mutant 2

PROTEIN SUBUNITS

Does not aggregate

Heterozygous diploid

Does not aggregate

Aggregates to form active enzyme

Aggregation of complementary subunits to form active enzyme

Figure 7.10 *Intragenic complementation. The peptides encoded by the two mutant genes have complementary tertiary structures and can aggregate in dissimilar pairs to produce an active enzyme*

3 Both complementing mutants must produce polypeptides serologically related to the normal polypeptide subunit, i.e. they must not be nonsense mutants (Section 8.4) or frameshift mutants (Section 9.2) which produce prematurely terminated polypeptides.

7.9 Supplementary Questions

7.1 What determines (**a**) the functional properties of a protein and (**b**) its specific three-dimensional pattern of folding?

7.2 Some amino acid substitutions within a protein abolish or greatly reduce its functional properties; other amino acid substitutions have little or no effect. Explain why this should be so.

7.3 List reasons why the one gene, one enzyme hypothesis is incorrect.

7.4 Why do phenylketonuriacs not suffer from a deficiency of tyrosine?

7.5 *trp-1*, *trp-2*, *trp-3* and *trp-4* are mutations in different genes of *Neurospora*, and each results in a requirement for tryptophan. These mutants respond as follows to presumed intermediates in the biosynthetic pathway for tryptophan:

trp-1 grows when provided with anthranilic acid, indole glycerol phosphate, indole, and tryptophan;

trp-2 responds to indole and tryptophan;

trp-3 responds to indole glycerol phosphate, indole and tryptophan; *trp-4* responds only to tryptophan.

What is the probable pathway for tryptophan biosynthesis?

7.6 The following table shows how three thiamine-requiring mutants of *Neurospora* respond to possible precursor compounds:-

	Growth in the presence of			
	minimal medium	thiazole	pyrimidine	thiamine
thi-1	–	–	–	+
thi-2	–	–	+	+
thi-3	–	+	–	+

Suggest a pathway for the biosynthesis of thiamine.

7.7 Tabulate some of the methods used for investigating complementation.

7.8 Three methionine-requiring mutants of *E. coli* are streaked as shown on minimal medium containing a trace of methionine. After 24 h there was dense growth at both ends of the *met-3* streak and at one end of the *met-1* streak:

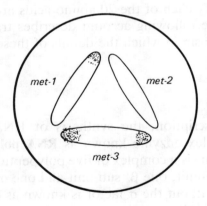

Suggest an explanation for this pattern of growth.

7.9 Why does intragenic complementation only occur between (**a**) alleles at certain gene loci and (**b**) particular pairs of alleles at these loci?

7.10 Why, when discussing dominance, is it more correct to refer to a dominant character than to a dominant allele?

8 From gene to protein

8.1 An Overview

Most structural genes function by encoding specific polypeptides (but some encode the RNA molecules involved in translation) and the first step in gene expression is for the gene to be **transcribed** onto a molecule of single-stranded RNA base complementary to one strand (the **antisense** strand) of the DNA duplex. Second, this **sense** strand of RNA, the messenger or mRNA, is **translated** by acting as a template against which amino acids are polymerised into a polypeptide. The exact sequence of amino acid residues in this polypeptide is determined by the sequence of nucleotides in the mRNA (and, hence, by the sequence of base pairs in the DNA) and the trinucleotide sequences which specify each of the 20 amino acids are known as the **genetic code** (Chapter 9).

The following account describes transcription and translation in *E. coli*, the organism in which the details of these processes were first revealed.

8.2 Transcription

Transcription, the synthesis of RNA against a DNA template, requires a complex enzyme known as **RNA polymerase** (or **RP**). The complete or **holoenzyme** is a complex of five polypeptide sub-units, two identical α sub-units, one β sub-unit, one β′ sub-unit and one σ sub-unit (or σ factor); the same enzyme, but without the σ factor is known as **core enzyme**.

A further requirement for accurate transcription is the presence on the DNA of two specific genetic control sequences; these are a **promoter**, the site at which transcription is initiated, and a **terminator**, at which transcription ceases; both are of the order of 40 bp long.

Transcription (Question 8.1) occurs in four stages:

(1) **RP binding**   The holoenzyme recognises the specific nucleotide sequence of the **promoter**, which precedes (i.e. lies **upstream** of) the gene or group of genes to be transcribed.

(2) **Initiation**   The RP holoenzyme unwinds the DNA within the promoter, recognises the antisense strand of the DNA as the strand to be transcribed and initiates mRNA synthesis a short distance upstream from the coding sequence of

the gene (Question 8.2). At this stage the sigma factor is no longer required, as its purpose seems to be the initial recognition of the promoter by the holo-enzyme; the sigma factor is released and recycles.

(3) **Elongation** The molecule of core RP moves along the DNA template and unwinds the DNA (Figure 8.1). At the same time the growing strand of nascent mRNA is extended in the 5′ to 3′ direction. As the RP moves further along the DNA, so the single-stranded DNA is rewound into a duplex molecule.

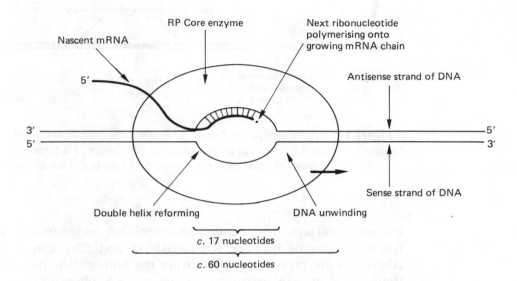

Figure 8.1 *Transcription. The molecule of RNA polymerase (RP) is moving from left to right and using the antisense DNA strand as a template for RNA synthesis*

(4) **Termination** Elongation continues until the entire gene or group of genes has been transcribed. Shortly beyond the end of the last gene (i.e. **downstream**) the RP encounters a **terminator** sequence; this signals the comple-tion of mRNA synthesis and causes the complex of RP, DNA and RNA to dissociate, releasing the newly synthesised mRNA (Question 8.3).

In bacteria most genes are transcribed in small groups producing a **polycis-tronic messenger** and the adjacent genes either may be contiguous or they may be separated by a short non-coding **intergenic region**. These relationships between DNA and mRNA are summarised in Figure 8.2.

Transcription in eucaryotes is described in Chapter 12. For the moment it is sufficient to note that (a) each gene is separately transcribed and (b) the primary transcripts are extensively processed before they move to the cytoplasm and act as templates for transcription.

8.3 The Translational Machinery

The four-letter language of the genetic code on the mRNA (A, U, C and G) must now be translated into the 20-amino-acid language of polypeptides. Since amino acids have no affinity for the bases on mRNA, a variety of **adaptor** molecules is required, each adaptor recognising a particular amino acid on the

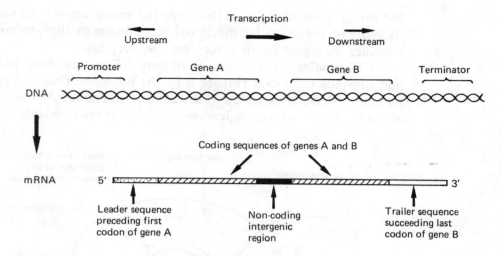

Figure 8.2 *A transcriptional unit. Transcription commences within a promoter just upstream from the first structural gene and terminates at a terminator just downstream from the last structural gene to be included on the mRNA*

one hand and a particular codon on the mRNA on the other. These adaptors are the molecules of **transfer RNA** (tRNA) and they can only function when ribosomes are present. The latter are the non-specific factories of the cell and they make protein according to the mRNA blueprint with which they are provided, using as raw materials the amino acids carried by the molecules of tRNA.

(a) Transfer RNA

Transfer RNA molecules are also gene transcripts but they are not translated. The primary transcripts are oversized and they have to be **processed** before they can assume their secondary and tertiary structures and function as molecules of tRNA; this involves (a) trimming the molecule to size and (b) chemical modification of certain of the bases so as to prevent them from forming base pairs by hydrogen bonding. This is important in maintaining the secondary structure of the molecule. Each tRNA is about 80 nucleotides long and is folded into a clover-leaf structure (Figure 8.3) held together by H bonding between complementary bases within the arms. This secondary structure is vital, and mutations which alter the nucleotide sequence may affect this structure and abolish or modify the activity of the tRNA.

Consider tRNA [Trp], which aligns the amino acid tryptophan against the mRNA template. This has two specificities. First, one arm contains the triplet CCA; this **anticodon** is base-complementary to the UGG codon and so the anticodon on the tRNA [Trp] can bind to the corresponding codon on the mRNA (Figure 8.4). Second, the free 3' end of the tRNA becomes **charged** with tryptophan.

In *E. coli* there are about 50 different tRNAs each transporting a specific amino acid and recognising specific codons on the mRNA.

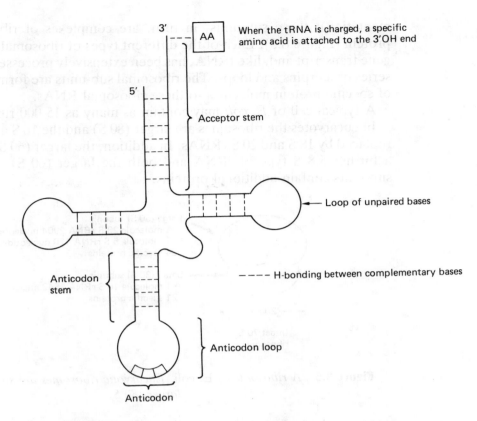

When the tRNA is charged, a specific amino acid is attached to the 3'OH end

AA

3'

5'

Acceptor stem

Loop of unpaired bases

Anticodon stem

- - - - H-bonding between complementary bases

Anticodon loop

Anticodon

Figure 8.3 *Transfer RNA. A typical tRNA molecule is between 75 and 85 nucleotides long and folded into a 'clover-leaf'-shaped structure. Many of the chemically modified bases are found in the loops and, since these 'unusual' bases are unable to form conventional base pairs, it is likely that one of their functions is to keep these regions single-stranded*

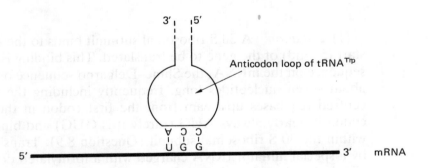

3'   5'

Anticodon loop of tRNA^Trp

A C C
U G G

5' ━━━━━━━━━━━━━━━━━━━━━━━━━━━ 3' mRNA

Figure 8.4 *Codon and anticodon. The pairing between the UGG codon and the CCA anticodon is antiparallel. Remember that sequences are always written in the 5' to 3' direction*

(b)   The Ribosomes

The ribosomes are the factory of the cell and each is formed by the association of a 30 S and a 50 S subunit (S, the Svedberg unit, is the coefficient of sedimentation); these sub-units, in turn, are complexes of ribosomal RNA and protein (Figure 8.5). Each of the different types of ribosomal RNA (rRNA) is a gene transcript and, like tRNA, has been extensively processed and folded into a series of hairpins and loops. The ribosomal sub-units are formed by the addition of specific protein molecules to these ribosomal RNAs.

A typical cell of *E. coli* may contain as many as 15 000 ribosomes.

In eucaryotes the ribosomes are larger (80 S) and the 16 S and 23 S rRNAs are replaced by 18 S and 20 S rRNAs. In addition, the larger (60 S) sub-unit contains a further 5.8 S type of rRNA and both the larger (60 S) and smaller (40 S) sub-units contain additional proteins.

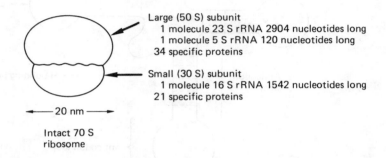

Figure 8.5 *A ribosome of* E. coli. *Eucaryotic ribosomes are similar but rather larger (80 S)*

8.4 Polypeptide Synthesis — Translation

Translation, the synthesis of a polypeptide using mRNA as a template (Questions 8.7, 8.8 and 8.9), occurs in three separate stages:

(1) *Initiation* A 30 S ribosomal sub-unit binds to the mRNA adjacent to the start (5′ end) of the gene to be translated. This binding is facilitated by a special sequence on the mRNA, the Shine–Delgarno sequence or **ribosome binding site**, about seven nucleotides long, frequently including the sequence AGGA and centred ten bases upstream from the first codon in the gene. This initiation codon is nearly always AUG (rarely it is GUG) and binding correctly aligns it within the 30 S ribosomal subunit (Question 8.9). Translation now commences by a special initiator tRNA charged with a formylated derivative of methionine aligning against the AUG initiating codon.

It is important to realise that (a) this initiator tRNA can ONLY respond to an AUG codon before the 50 S subunit is added; thus, internal AUG codons are recognised normally by a tRNA charged with methionine, (b) no other tRNA can recognise the mRNA–30 S complex and (c) in procaryotes either the formyl group is cleaved from the formyl-methionine residue at the completion of polypeptide synthesis or the entire formyl-methionine residue is cleaved, leaving the next amino acid in sequence as the N-terminal residue.

(2) *Elongation* The ribosome is now completed by the addition of a 50 S sub-unit and moves along the mRNA in the 5′ to 3′ direction. As it processes

along the mRNA, each successive mRNA codon is 'exposed' within the ribosome and is recognised by and transiently H bonded to the corresponding anticodon on the correct molecule of charged tRNA. The amino acid on this charged tRNA is simultaneously released from its tRNA and joined onto the carboxy end of the growing polypeptide chain by the formation of a peptide bond. This process continues until the end of the gene is reached.

(3) *Termination* Each of the three chain termination triplets, UAA, UAG and UGA, signals the end of translation. These triplets are not recognised by any normal tRNA, and, whenever one is encountered by a ribosome, the mRNA and the completed polypeptide are released and the ribosome dissociates into its 50 S and 30 S sub units, which then recycle.

When one or other of these chain termination triplets is generated by mutation in the middle of a gene, it causes the premature termination of polypeptide synthesis, so that only an N-terminal polypeptide fragment is produced; such mutations are referred to as **nonsense mutations**.

Each gene on a molecule of mRNA is separately translated and the amino acids are added at the rate of about 15 per second at 37 ° C.

8.5 The Coupling of Transcription and Translation

A special feature of bacterial systems is that transcription and translation are coupled; as RNA polymerase is transcribing messenger, so ribosomes are binding to the messenger and initiating translation. Many ribosomes (up to 50, or 1 every 80 nucleotides) may be simultaneously translating a single molecule of nascent mRNA and each complex of messenger plus attached ribosomes is a **polysome** (Figure 8.6).

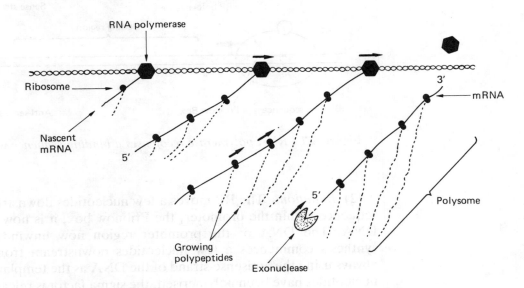

Figure 8.6 *The coupling of transcription and translation in* E. coli. *As molecules of RP are transcribing mRNA, so ribosomes are translating these newly synthesised messengers. The entire complex of DNA, RP, mRNA, ribosomes and polypeptides is known as a polysome*

Furthermore, bacterial messengers are unstable and may have a half-life of only $1\frac{1}{2}$ min; this means that as RNA polymerase is transcribing mRNA at its 3' end, so nucleases are digesting it from the 5' end.

In eucaryotes this coupling is not possible, because the ribosomes lie outside the nucleus, either lining the outer surface of the endoplasmic reticulum or free in the cytoplasm; hence, transcription occurs in the nucleus but the messenger must move to the cytoplasm before translation can commence.

8.6 Questions and Answers

Question 8.1

Outline the process of transcription in a bacterial cell.

Answer 8.1

Transcription is the production of single-stranded RNA against a DNA template by the enzyme RNA polymerase (RP). The steps in this process are:

(1) *RP binding* A molecule of RP holoenzyme binds loosely to the DNA at a promoter sequence located just upstream from the coding sequence(s) to be transcribed (Figure 8.7).

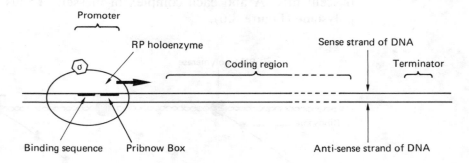

Figure 8.7 *RNA polymerase recognises a binding sequence within the promoter*

(2) *Initiation* The RP moves a few nucleotides downstream to another short sequence within the promoter, the Pribnow box; it is now tightly bound to the DNA. The DNA of the promoter region now unwinds (melts) and RNA synthesis commences a few nucleotides downstream from the Pribnow box, always using the antisense strand of the DNA as the template. After the first few nucleotides have been polymerised, the sigma factor is released and it recycles — it is no longer required (Figure 8.8).

(3) *Elongation* The molecule of core RP moves along the DNA template, unwinds the DNA and polymerises the nucleoside triphosphate that is complementary to the next exposed base on the DNA template strand. As the RP

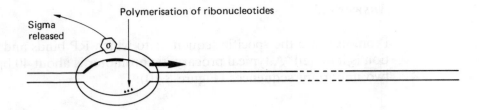

Figure 8.8 *The initiation of transcription. RNA polymerase moves a few nucleotides downstream to the Pribnow box; the DNA within the molecule of RP melts and polymerisation commences*

progresses along the DNA, the RNA is released from the template strand and the DNA double helix is reformed (Figure 8.9).

(4) *Termination* When all the coding sequences have been transcribed, the RP reaches a termination sequence, the terminator. The transcriptional complex dissociates and the RP and the newly transcribed RNA are released from the template.

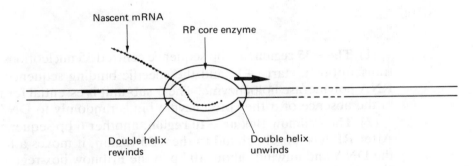

Figure 8.9 *Elongation of the transcript. The sigma factor is released and the core enzyme moves downstream, adding ribonucleotides to the growing molecule of nascent mRNA*

NOTES

1 Nucleoside triphosphates are always polymerised in the 5′ to 3′ direction.
2 The 5′ terminal nucleotide is always a ribonucleoside triphosphate.

Question 8.2

Outline the structure and function of a typical procaryotic promoter. Explain what is meant by a consensus sequence.

Answer 8.2

Promoters are the specific sequences to which RP binds and at which transcription is initiated. A typical procaryotic promoter is about 40 bp long and contains two important sequences (Figure 8.10):

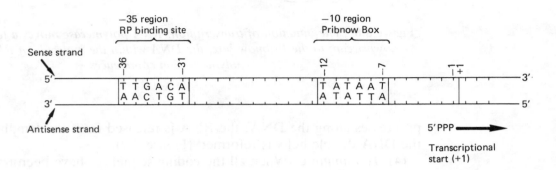

Figure 8.10 *A procaryotic promoter. The two important sequences are the site to which RP initially binds (the −35 or binding site) and the sequence which aligns RP against the antisense strand of the DNA (the −10 region or Pribnow box) and ensures accurate initiation*

(1) **The −35 region**, a 6 bp sequence centred 35 nucleotides upstream from the transcriptional start (+1) and the specific binding sequence recognised by the RNA polymerase holoenzyme. The σ sub-unit is essential for this recognition, as in the absence of σ the core enzyme binds randomly to DNA.

(2) **The Pribnow Box or −10 region**, another 6 bp sequence centred on −10. After RP has loosely bound to the −35 region, it moves a few base pairs along the DNA and unwinds about 10 bp in the Pribnow box region, forming a tightly bound RP–DNA complex. It is thought that the Pribnow box sequence enables RP to recognise the antisense strand of the DNA duplex, so ensuring that the correct DNA strand is transcribed and in the correct direction. The RP now initiates RNA synthesis by inserting a ribonucleoside triphosphate opposite the base in the +1 position on the antisense DNA strand.

The Pribnow box has the consensus sequence

5′ TATAAT 3′   sense strand
3′ ATATTA 5′   anti-sense strand

usually written simply as TATAAT. By consensus sequence we mean that this sequence or a very closely related sequence is found at this position in all promoters; only in a very few promoters does the Pribnow box differ from the consensus sequence by more than one or two nucleotides. Similarly, the −35 region has the consensus sequence TTGACA.

NOTE

In general, the closer the actual sequences are to the consensus sequences the more efficient is that promoter. Thus, mutations which remove a match with the consensus sequence usually result in reduced promoter activity (these are **down-promoter mutations**), while mutations which increase the match result in more efficient promoters (**up-promoter mutations**).

Question 8.3

How is transcription terminated at the end of a gene or group of genes?

Answer 8.3

RNA synthesis ceases at specific nucleotide sequences called **terminators**, present on the DNA template and the RNA transcript. The two special sequences present in a typical terminator (Figure 8.11) are best considered by examining the RNA transcript (which is base-complementary to the antisense strand of the DNA). These are:

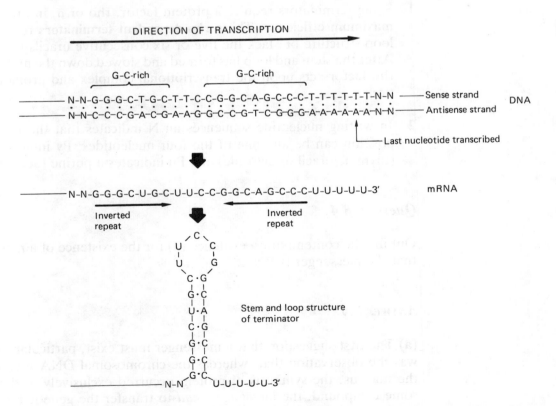

Figure 8.11 *A typical rho-independent terminator. The terminator sequence on the mRNA forms a stem and loop structure which slows down the molecule of RP; this allows the weak rU–dA base pairs to dissociate, releasing the newly synthesised mRNA*

139

(1) a G–C-rich inverted repeat sequence which allows the newly transcribed RNA to form a stable stem and loop structure.

(2) 5–6 uridine residues at the 3′ terminus of the RNA. Note that transcription has terminated within the corresponding sequence of A–T base pairs in the DNA.

It is thought that the terminator acts as follows:

(1) As soon as the molecule of RNA polymerase core enzyme (RP) reaches a terminator sequence, it slows down; this is because the terminator is GC-rich and G–C base pairs are transcribed more slowly than are A–T base pairs.

(2) Immediately the terminator has been transcribed, the stem and loop structure forms within the DNA–RNA–RP complex; this acts as a brake and hinders further forward movement of the molecule of RP.

(3) At this stage the newly synthesised RNA is bonded to the DNA template strand only by five or six very weak rU–dA bonds; these rU–dA base pairs are 200 times less stable than any other ribonucleotide–deoxyribonucleotide base pair and this allows the weakly bonded RNA–DNA hybrid region to dissociate, releasing free mRNA, DNA and RP.

NOTES

1 Some terminators require a protein factor, rho or ρ, in order to operate at maximum efficiency. These **rho-dependent terminators** retain the stem and loop structure but lack the five or six consecutive uracils in the mRNA tail. After the stem and loop has formed and slowed down the molecule of RP, the rho factor acts upon the transcriptional complex and promotes its dissociation.

2 In writing nucleotide sequences an N indicates that the nucleotide at that position can be any one of the four nucleotides; Py indicates a pyrimidine (thymine/uracil or cytosine) and Pu indicates a purine (adenine or guanine).

Question 8.4

Outline the contemporary evidence (a) for the existence of a messenger and (b) that the messenger is RNA.

Answer 8.4

(a) The first suggestion that a messenger must exist, particularly in eucaryotes, was the observation that, whereas the chromosomal DNA was located only in the nucleus, the synthesis of protein occurred exclusively in the cytoplasm — some compound, the messenger, *had* to transfer the genetic message from the nucleus to the cytoplasm. The first evidence suggesting that this messenger was RNA came in 1957, when Elliot Volkin and Lazarus Astrachan noted that immediately after the infection of *E. coli* cells by phage T2 there is (1) an immediate halt to the synthesis of bacterial RNA and protein, followed by (2) the rapid synthesis of T2-specific RNA and protein; furthermore, this RNA had base ratios that reflected the base ratios of T2 DNA rather than those of bacterial DNA.

(b) More convincing evidence for the messenger RNA hypothesis came from hybridisation experiments carried out in 1961 by Bernard Hall and Sol Spiegelman. They isolated the RNA (the putative messenger) that appears in the cells of *E. coli* immediately after infection by phage T2 and they mixed this single-stranded RNA with the single-stranded DNA formed by gently heating (denaturing) either *E. coli* or phage T2 DNA. When this mixture of RNA and DNA is gently cooled (1) double-stranded DNA molecules reform and (2) DNA–RNA hybrid duplex molecules form when the single-stranded DNA and RNA molecules are base-complementary. They found that the RNA appearing after infection by T2 formed DNA–RNA hybrid duplex molecules with the DNA extracted from T2 phages but NOT with the DNA extracted from *E. coli*. Thus, the RNA appearing in the phage-infected cells was base-complementary to at least one of the strands of T2 duplex DNA.

Question 8.5

What evidence suggests that, for a given gene, only one strand of the template DNA is transcribed?

Answer 8.5

If both strands of a gene are transcribed, then, since RNA polymerase only operates in a 5′ to 3′ direction, the two strands of the DNA duplex would be transcribed in opposite (physical) directions and the two messengers would carry nucleotide sequences that were both inverted and base-complementary to each other; each could encode a different polypeptide. Although certain DNA sequences are transcribed in both directions, this is unlikely to be the normal situation.

The first definitive evidence came from studies using phage SP8, which infects *Bacillus subtilis*. The DNA of SP8 is unusual, as one of the strands is purine-rich (the 'heavy' strand) and the other is pyrimidine-rich (the 'light' strand); if this duplex DNA is gently heated, it separates into its two component strands (i.e. the DNA is denatured) and these strands can then be separated by density gradient centrifugation.

In 1963 Julius Marmur and Paul Doty extracted the RNA appearing in *B. subtilis* cells following SP8 infection and found that, whereas it would hybridise (i.e. form DNA–RNA hybrid molecules by complementary base pairing) to the separated 'heavy' strands of the phage DNA, it would not hybridise with the 'light' strands. Clearly, the messenger was base-complementary to only one of the DNA strands (the 'heavy' strand) and so only one of the two DNA strands had acted as a template for transcription (Figure 8.12).

Note that Marmur and Doty were lucky in choosing SP8, as all the phage genes are transcribed from the same strand of DNA. With other phages,

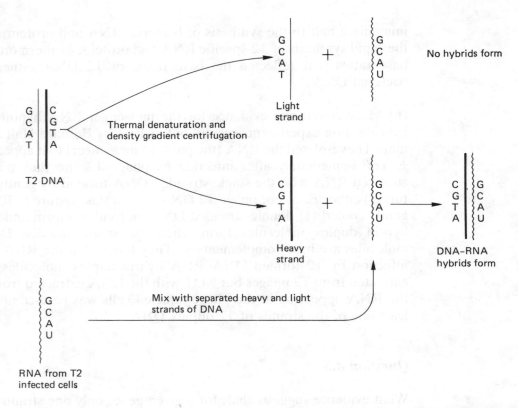

Figure 8.12 *The Marmur and Doty experiment. The RNA extracted from T2-infected cells will only hybridise with the 'heavy' (purine-rich) strand of T2 DNA*

including T4 and lambda, some genes are transcribed from one strand and other genes from the alternative strand; thus, this experiment would not have worked had one of these phages been used instead of SP8.

Question 8.6

How would you **(a)** show that RNA is synthesised in the 5′ to 3′ direction, **(b)** distinguish between a fragment of RNA transcribed from the initiation sequence within a promoter and a fragment whose 5′ terminal sequence had been removed by processing, and **(c)** show that polypeptide synthesis occurs in the amino-terminal to the carboxy-terminal direction?

Answer 8.6

(a) *E. coli* is grown at O°; this reduces the rate of transcription by a factor of 40, so that only one nucleotide is added every 13 seconds. The cells are then fed with [14C]-uridine, which specifically labels the newly synthesised RNA. When the growing RNA chains are extracted and analysed, it is found that the 14C label first appears at the 3′ end of the growing molecules; this identifies the end to which new nucleotides are added so that transcription **must** occur in the 5′ to 3′ direction.

(b) During transcription it is the 5′ nucleoside triphosphates that are polymerised onto the 3′–OH end of the growing RNA chains; this involves the release of pyrophosphate (i.e. two phosphate groups) and the formation of a phosphodiester bond. Thus, only the initiating nucleoside will have three phosphate groups attached at the 5′ end; if the RNA has been processed, then the terminal 5′ nucleoside triphosphate will have been removed, leaving a terminal nucleoside monophosphate.

(c) Cells of *E. coli* are fed with ^{14}C-labelled leucine (or any other labelled amino acid). This will be incorporated ONLY at the growing ends of polypeptide chains. When the polypeptides are extracted and examined, the label will identify the last regions synthesised. It is found that the highest concentration of label always appears at the carboxy termini and the lowest at the amino termini.

Question 8.7

Compare the transcription and processing of rRNA and tRNA in *E. coli*.

Answer 8.7

In *E. coli* there are seven sets of closely linked *rrn* genes enabling the large amounts of rRNA required by the cell to be transcribed. Each transcription unit contains one 16 S, one 23 S and one 5 S rRNA gene and, in addition, has a gene for one or other of the tRNAs between the 16 S and 23 S genes; there may also be one or two further tRNA genes beyond the 5 S gene (Figure 8.13.i). The remaining 50 or so tRNA genes are located singly or in small groups elsewhere in the genome.

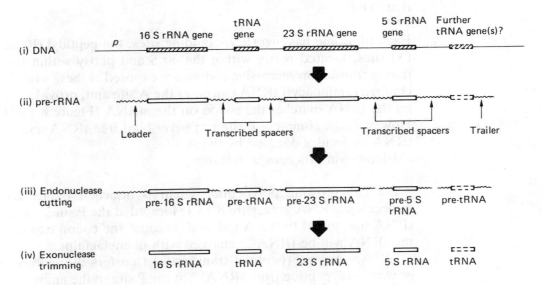

Figure 8.13 *The transcription and processing of rRNA and tRNA in* E. coli. *In addition to being trimmed to size by processing, some of the nucleosides are chemically modified by, for example, methylation or deamination*

Each complex transcriptional unit is transcribed (Figure 8.13.ii) as a **precursor rRNA (pre-rRNA)** molecule over 5500 nucleotides long or as a molecule of **precursor tRNA (pre-tRNA)**. These precursors must now be **processed**.

The first step in processing is the activity of endonucleases which cut the precursor molecules. Thus, the pre-rRNA is cut into pre-16 S, pre-23 S and pre-5 S rRNA and pre-tRNA (Figure 8.13.iii). Secondary processing is carried out by exonucleases which trim the ends of the precursors to form 16 S, 23 S and 5 S rRNA and tRNA. The independently transcribed pre-tRNA molecules are similarly cut and trimmed to size.

During processing **modification** also takes place. The 16 S and 23 S rRNA are modified by the chemical addition of methyl groups to certain nucleosides, but the modification of the tRNAs is more complex.

In tRNAs between 10% and 17% of the nucleosides are modified by, for example, methylation or deamination. There are over 50 different modified or **minor bases**, and they are all unable to form conventional base pairs; this appears to be part of their function, as they are found principally in the unpaired loops of the clover-leaf structure.

All tRNA molecules end with the sequence 5′ … CCA–OH3′; this sequence is vital, as the terminal A is the site to which the amino acid will be attached. In some instances this sequence is part of the primary transcript, but in others it is added enzymatically after the completion of processing.

Question 8.8

How are successive amino acids added to a polypeptide chain during translation?

Answer 8.8

Each ribosome has *two* tRNA binding sites, the **peptidyl (P)** and the **aminoacyl (A)** sites, located partly within the 30 S and partly within the 50 S sub-units; during translation successive codons are exposed at these sites. Free molecules of charged (aminoacyl) tRNA can enter the A site and, provided that the anticodon on the tRNA matches the codon on the mRNA (Figure 8.14.1), will bind to it. However, the complete P site can *never* bind free tRNA and it will only accept tRNA molecules donated by the A site.

Elongation proceeds as follows:

(1) The growing polypeptide chain is attached by its carboxy end to a molecule of tRNA^{Tyr} (Figure 8.14.1) located at the P site; a molecule of charged tRNA has bound to the A site and, because the codon exposed there is UUU, the tRNA will be tRNA^{Phe} charged with phenylalanine.

(2) An enzyme (peptidyl transferase) transfers the carboxy terminal of the growing polypeptide from tRNA^{Tyr} at the P site to the amino group of the newly accepted phenylalanyl tRNA. This creates a new peptide bond and releases the now uncharged (peptidyl) tRNA^{Tyr} from the P site (Figure 8.14.2).

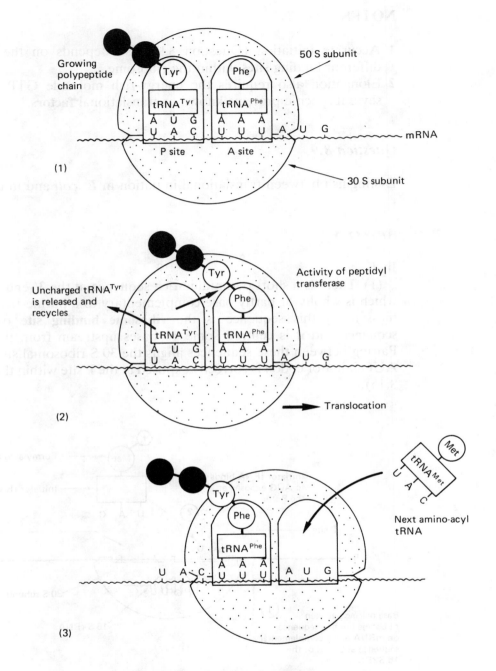

Figure 8.14 *Polypeptide chain elongation*

(3) The ribosome now translocates (i.e. moves) exactly one codon along the mRNA, so that tRNA^{Phe}, previously at the A site, is now located at the P site (Figure 8.14.3). The next codon in sequence (AUG) is now exposed at the A site, which can now accept a molecule of charged tRNA^{Met}.

This process is repeated many times as the ribosome translocates along the messenger and continues until the end of the gene is reached.

NOTES

1 Accurate initiation (Question 8.9) also depends on the presence of two different binding sites within the ribosome.
2 Elongation also requires the energy-rich molecule GTP, cyclic AMP and several specific proteins known as translational **factors**.

Question 8.9

Distinguish between translational initiation in *E. coli* and in eucaryotes.

Answer 8.9

In *E. coli:*

(1) There is a short pyrimidine-rich sequence at the 3' end of the 16 S rRNA which is wholly or partly base-complementary to a purine-rich sequence on the messenger; this sequence is the **ribosome binding site** or **Shine–Delgarno sequence** and it is centred some 10 bases upstream from the AUG initiator. Pairing between these sequences aligns the 30 S ribosomal sub-unit, so that the AUG initiator codon is correctly exposed at the P site within the sub-unit (Figure 8.15).

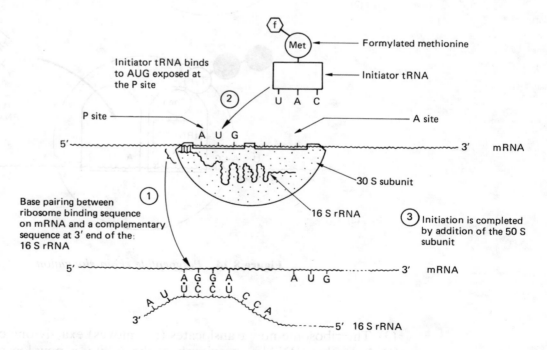

Figure 8.15 *The initiation of translation in* E. coli. *The 30 S ribosomal sub-unit binds to the mRNA by complementary base pairing between a sequence at the 3' end of the 16 S rRNA and the ribosome binding sequence on the mRNA. This aligns the AUG initiation codon within the partial P site, where it is recognised by a special initiator tRNA charged with formyl-methionine*

(2) This AUG in the 30 S–mRNA complex is recognised by a special initiator tRNA charged with a formylated methionine.

(3) The 50 S sub-unit is added and elongation commences.

In eucaryotes there is no equivalent of a ribosome binding sequence on the messenger, nor is there any special initiator tRNA. However, all eucaryotic messengers terminate at their 5′ ends with a 7-methyl guanosine, known as the **cap** (see Section 12.3d).

In eucaryotes:

(1) A 40 S ribosomal sub-unit recognises the methylated cap on the messenger and binds to its 5′ end (Figure 8.16).

(2) The 40 S subunit migrates along the mRNA until it comes to the first AUG sequence; this AUG is now exposed at the P site within the 40 S sub-unit.

(3) This AUG codon is recognised by tRNA^{Met}, the *same* tRNA that recognises internal AUG codons.

(4) The 60 S sub-unit is added and elongation commences.

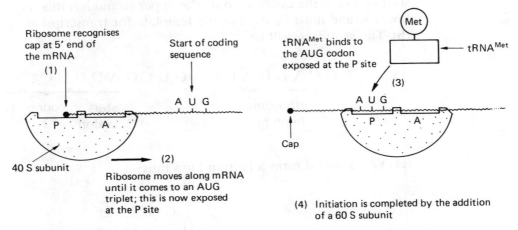

Figure 8.16 *The initiation of translation in eucaryotes. A 40 S ribosomal sub-unit recognises the 5′ cap and moves along the mRNA until it is aligned with the first AUG codon it encounters. This initiation codon is now recognised by the normal tRNA^{Met}*

NOTES

1 In *E. coli* the 3′ end of the 16 S rRNA has the sequence 5′ . . . ACCUC-CUUA3′, while the ribosome binding site always includes a sequence such as 5′AGGA3′ or 5′GAGG3′.

2 Initiation is the only time that a charged tRNA can enter and bind to the P site — and then *only* before the larger ribosomal sub-unit is added.

Question 8.10

You have isolated a fragment of DNA thought to include the first few codons of a gene encoding a polypeptide. This fragment has the sequence

5′ C G C A G G A T C A G T C G A T G T C C T G T G
3′ G C G T C C T A G T C A G C T A C T G G A C A C

(a) Which strand must be the template for transcription?
(b) What is the nucleotide sequence of the mRNA?
(c) Could this mRNA form into a secondary structure?
(d) Where does translation commence and in which direction does it proceed?
(e) Is the DNA more likely to have been isolated from a procaryote or a eucaryote?

Answer 8.10

(a) If this sequence includes the start of a gene encoding a polypeptide, then there must be an ATG sequence on one strand (this corresponds to an AUG start codon in the mRNA); only the upper strand has this sequence. Hence, the lower strand must be used as the template for transcription.

(b) The sequence will be

5′ . . . C G C A G G A U C A G U C G A U G U C C U G U G . . . 3′

| ribosome binding | | start codon | codon 2 | codon 3 |

(c) Yes, it could form a hairpin loop thus:

```
              U  C
         G        G
      A             A
      C             U
       U           G

         A–U
         G–C
         G–C
         A–U
         C–G
      C  G    U  G
```

(d) Translation commences at the AUG initiation codon and proceeds in the 5′ to 3′ direction (left to right).

(e) There is an AGGA sequence centred nine nucleotides upstream from the initiation codon. This is complementary to the UCCU of the Shine–Delgarno sequence (see Figure 8.15) and is the putative ribosome binding site. These sites are absent in eucaryotes, which suggests that the DNA is procaryotic in origin.

Question 8.11

Write on 'wobble'.

Answer 8.11

The genetic code is highly degenerate and, with the exception of methionine and tryptophan, all the amino acids are specified by more than one codon. After the code had been deciphered (Chapter 9), it was initially thought that each of the 61 codons (Table 9.1) was recognised by a unique species of tRNA, each with its own anticodon. However, it soon became apparent from the triplet binding experiments (Section 9.3d) that some species of tRNA could recognise more than one codon — thus, one species of tRNA^{Val} could bind to both GUA and GUG but not to GUC and GUU.

In 1966 Crick was struck by the pattern of degeneracy whereby a change in the first or second base of a codon generally produces a codon for a different amino acid, whereas a change in the third position frequently (but not always) generates a different codon for the same amino acid. Thus, UCU, UCC, UCA and UCG are all code words for serine but CCU, CCC, CCA and CCG are code words for proline.

To explain this Crick proposed the **Wobble Hypothesis**. This states that in the third position of the codon there is some non-standard pairing, or wobble, with the corresponding base in the first position of the anticodon — this enables an anticodon to recognise more than one codon. According to this hypothesis,

a U at the 5′ end of an anticodon can pair with an A (normally) or G (with wobble) at the 3′ end of the codon;

a G at the 5′ end of an anticodon can pair with either C (normal) or U (with wobble) at the 3′ end of the codon;

an I (inosine, a derivative of guanine) at the 5′ end of an anticodon can pair with C (normal) or with either U or A (with wobble), at the 3′ end of the codon;

a C or A at the 5′ end of the anticodon does not show wobble and can only pair normally with G or U, respectively.

Thus, the anticodon UAC will recognise the codons GUA (normal) and GUG (with wobble), both codons for valine. However, GUC and GUU are also codons for valine and these must be read by a different species of tRNA^{Val}. This tRNA has the anticodon IAC, which can recognise not only GUC and GUU, but also GUA; hence, the GUA codon can be read by either species of tRNA^{Val}.

8.7 Supplementary Questions

8.1 Why is it that only one strand of the DNA duplex is normally transcribed?

8.2 What three sequences are essential for the accurate transcription of a procaryotic messenger?

8.3 Suggest how differences in RP–promoter interactions can result in the transcription of different genes.

8.4 In procaryotes the ribosomal and transfer RNAs are relatively stable and long-lived, whereas the messengers are unstable and short-lived. Suggest an explanation for this difference in stability.

8.5 How is it that the rRNA and tRNA molecules are so much more stable than the messengers?

8.6 What two special properties distinguish initiator tRNA from all the other tRNAs?

8.7 Assuming the validity of the wobble hypothesis, what is the minimum number of tRNA species required to translate all 61 amino acid codons (you may use the table of the genetic code set out in Table 9.1)?

8.8 A molecule of RNA has the sequence

$$5' \text{ A A U G A G U A G C A U C G G C U A C C G A G G } 3'$$

After digestion with a nuclease specific for single-stranded RNA, you are left with a double-stranded fragment with the sequence

$$5' \text{ G U A G C } 3'$$
$$3' \text{ C A U C G } 5'$$

What was the structure of the original molecule?

8.9 Distinguish between the roles of tRNA and mRNA during translation.

8.10 How is an amino acid joined to the correct molecule of tRNA?

9 The genetic code

9.1 Introduction

Most structural genes act by specifying the primary structures of specific protein molecules and, since the only difference between one gene and another lies in their nucleotide sequences, it follows that the particular nucleotide sequence of a gene must determine the amino acid sequence of the protein encoded by that gene. This is the **sequence hypothesis**, first stated by Francis Crick in 1958. The way this information is stored is referred to as the **genetic code**.

Since mRNA is identical in sequence with one strand of the DNA and base-complementary to the other strand (with U replacing T), it is more convenient to write coding sequences as they occur along the mRNA — thus, the four letters A, U, C and G are the genetic code, rather than the base pairs $\frac{A T C}{T A G}$ and $\frac{G}{C}$.

From the outset it was clear that the code word, or **codon**, for each amino acid must contain at least three letters: there are 20 amino acids commonly incorporated into protein and a two-letter code could encode, at the most, 16 amino acids (i.e. 4×4 doublets, AA AC AG AU CA CC..., etc.), whereas a three-letter or **triplet** code can spell $4 \times 4 \times 4$ or 64 different codons (AAA AAT ATC GTG ..., etc.).

9.2 The General Nature of the Genetic Code

This, the theoretical basis of the code, was first established in 1961 as the result of some very elegant experiments carried out by Francis Crick and Sydney Brenner, using T4, a virulent phage that infects *E. coli*.

Their experiments established that:

(1) The code is **triplet** code.
(2) It is **non-overlapping**.
(3) It is **commaless**.
(4) Successive codons in the message are **read from a fixed starting point**.

Thus, the message

5′ A G U A U G C A U U C A G C U A G C G A C . . . . 3′

151

is always read

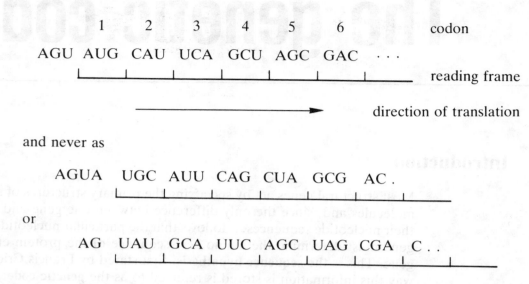

and never as

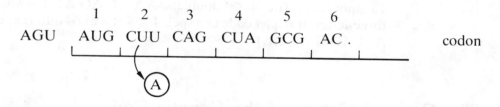

or

$$\text{AG} \quad \text{UAU} \quad \text{GCA} \quad \text{UUC} \quad \text{AGC} \quad \text{UAG} \quad \text{CGA} \quad \text{C} \cdot \cdot$$

Furthermore, because the code is **commaless** (i.e. it is unpunctuated and successive codons are not separated from each other by a short untranslated sequence), the deletion (or addition) of a single base pair from within the coding sequence of a tract of DNA will result in the loss (or addition) of a single nucleotide from a codon in the mRNA: this, in effect, will put the reading frame out of alignment, so that all downstream codons are incorrectly read. For example, if mutation deletes an A from codon 2 (arrowed in the above sequence), then the message will be read as

$$\begin{array}{ccccccc} 1 & 2 & 3 & 4 & 5 & 6 & \\ \text{AGU} & \text{AUG} & \text{CUU} & \text{CAG} & \text{CUA} & \text{GCG} & \text{AC} \cdot \end{array} \qquad \text{codon}$$

i.e. all the codons from and including codon 2 will be misread and the protein will contain a hopelessly wrong sequence of amino acids.

Mutations which add or delete a base pair from the DNA and so alter the phase in which the genetic code is read are referred to as **frameshift mutations**.

9.3 Deciphering the Code — The Biochemistry of the Genetic Code

Once the theoretical basis of the code had been established, it was necessary to determine the composition of the triplet or triplets encoding each amino acid. This work, carried out between 1961 and 1965, largely by two groups independently led by Matthew Nirenberg and Severo Ochoa, was made possible by the discovery of a system for the *in vitro* synthesis of proteins.

(a) *In vitro* Protein Synthesis

Extracts of *E. coli* cells contain ribosomes, transfer RNA, messenger RNA and a variety of enzymes — all the components necessary for protein synthesis. If such extracts are mixed with (1) energy-rich compounds such as ATP and GPT and (2) the 20 amino acids, then a small amount of protein synthesis takes place. If one of the amino acids used is labelled with ^{14}C, then the amount of protein synthesis is shown by the amount of the ^{14}C label incorporated into protein. In such a system protein synthesis stops after 30–60 mins, because the mRNA has a short half-life and is all degraded by this time. However, protein synthesis is resumed if an enzymatically synthesised RNA is now added to this system. The experiments of Nirenberg and Ochoa used, at first, artificial messengers of known composition (i.e. the ratio A:T:C:G was known but not the sequence) and later RNAs of known sequence. For each synthetic messenger it was determined which amino acids were incorporated into protein; this was determined by carrying out 20 replica experiments, and in each a different amino acid was labelled.

(b) The Use of Homopolymers

The first synthetic messenger used was a **homopolymer** of uracil (i.e. a repeating sequence of uracil ribonucleotides) known as polyU, and it was found that this *only* stimulated the incorporation of phenylalanine into protein: thus, UUU was identified as a code word for phenylalanine. Similarly, by using polyA and polyC they soon discovered that AAA was a codon for lysine and CCC a code word for lysine.

(c) The Use of Random Heteropolymers

The **heteropolymers** used contained two or more ribonucleotides in a known ratio but in a random sequence, and by comparing the relative frequencies with which different heteropolymers stimulated the incorporation of particular amino acids into protein it was possible to deduce the composition (but not the sequence) of many of the codons. For example, when a polyUC copolymer containing 5U:1C was used as messenger, four amino acids were incorporated at relative frequencies of

phenylalanine 100   serine 37   proline 12   leucine 5

(expressing the frequencies of incorporation as percentages of the total amount of phenylalanine incorporated).

In polyUC 5U:1C the most frequent triplet is UUU, occurring with a frequency of $5/6 \times 5/6 \times 5/6$ or 125/216. The next most common triplets contain 2U and 1C (i.e. UUC, UCU and CUU) and *each* occurs with a frequency of $5/6 \times 5/6 \times 1/6$ or 25/216, while the 1U + 2C triplets (UCC, CUC and CCU) occur with a frequency of $5/6 \times 1/6 \times 1/6$ or 5/216 each: the least common triplet, CCC (which we know encodes proline) occurs at a frequency of $1/6 \times 1/6 \times 1/6$ or 1/216. Thus, the relative frequencies of the four groups of codons are 125:25:5:1 or 100:20:4:0.8. The most frequent codon is UUU and we know this encodes phenylalanine; the next most common codons contain 2U + 1C and, therefore, it is most probable that serine, the next most commonly incorporated amino acid, is encoded by one of these triplets. We know that proline is encoded by CCC, the least common triplet (0.8), and its high frequency of incorporation (12%) suggests that at least one of the 1U + 2C codons also encodes proline.

Further experiments showed that both polyUC and polyUG stimulated the incorporation of leucine, so there must be at least two codons for leucine – one containing U and C and the other U and G: this was a clear demonstration that the code was **degenerate** and that some amino acids were encoded by at least two different codons.

(d)   Triplet Binding Experiments

In 1964 two new advances led to the rapid deciphering of the code. The first, developed by Nirenberg and his co-workers, was based on the fact that each species of tRNA has a specific anticodon which, in the presence of ribosomes, binds to base-complementary codons on the mRNA. By this time it was possible to synthesise specific trinucleotides (i.e. triplets) and Nirenberg added these to his *in vitro* system (Section 9.3a) instead of a heteropolymer: he found that although they did not result in the incorporation of amino acids into protein, each specific triplet caused specific tRNA molecules, together with their attached amino acids, to bind to the ribosomes. After incubation the mixture was passed through a nitrocellulose filter which retained only the ribosomes and any bound tRNA–amino acid complexes. In 20 replica experiments a different amino acid was ^{14}C-labelled and which amino acid was bound by a specific triplet was revealed by the presence of the ^{14}C label on the filter.

This enabled the *sequence* of nearly 50 codons to be determined. For example, the use of random heteropolymers had shown that the codons for valine, cystine and leucine probably contained 2U + 1G but these experiments showed that GUU triplets bound (and, hence, was the code word for) valine, UGU bound cystine and UUG bound leucine.

(e)   The Use of Regular Copolymers

The second advance was pioneered by Gobind Khorana; he had succeeded in synthesising long RNA molecules with regularly repeating sequences of bases, and he used these in an *in vitro* system and then determined the amino acid sequence of the resulting polypeptides.

One copolymer he used was poly(UC) — a regularly repeating series of UC dinucleotides, UCUCUCUCUCUCUCUC …. This stimulated the production of a polypeptide with alternating leucine and serine residues. This is expected with triplet code, since the message will be read

UCU CUC UCU CUC … … …

i.e. as alternating UCU and CUC codons; thus, one of these codons is the code word for leucine and the other is the code word for serine. Other regular copolymers used included poly(AC), poly(AG), poly(GUA) and poly(UAUC).

Note that poly(UC) indicates a regular copolymer, while polyUC indicates a copolymer with randomly distributed U and C nucleotides.

9.4 Initiation and Termination Codons

By 1965 61 of the 64 codons had been assigned to particular amino acids; the remaining three codons do not encode amino acids but are the chain termination or nonsense codons, UAG (amber), UAA (ochre) and UGA (opal). These occur naturally at the ends of genes and they signal the termination of polypeptide synthesis (Section 8.4). However, if mutation occurs, it may generate a nonsense codon *within* a gene and this results in the premature termination of translation and the production of an N-terminal polypeptide fragment. Furthermore, the AUG codon has a dual role: when it occurs within a gene, it is located within an intact ribosome–mRNA complex and is read normally as methionine, but when it occurs at the start of a gene, it lies within the 30 S rRNA–mRNA initiation complex and is recognised by a special initiator tRNA charged with *N*-formylmethionine. Thus, AUG (and sometimes GUG) also acts as the initiation codon (Section 8.4).

9.5 Confirmation of the Genetic Code

The most direct way to decipher the code would have been to compare the nucleotide sequence of a gene or its mRNA with the amino acid sequence of the polypeptide it encodes. However, this was not possible in the early 1960s, and was not achieved until 1972 as a result of the work of Walter Fiers and his colleagues on phage MS2. MS2 is a very small phage infecting F^+ and Hfr strains of *E. coli*, and its chromosome is a molecule of RNA only 3569 nucleotides long and containing just four genes. By 1972 Fiers had sequenced both the coat protein gene and the coat protein itself, making it possible to associate directly particular codons with particular amino acids. These experiments were the most direct and most exact confirmation of the genetic code.

They also found that *in vivo* phage MS2 uses all 61 of the amino acid codons (confirming that the code is degenerate, that some amino acids are encoded by several different codons) and all three chain termination codons.

The genetic code is set out in Table 9.1.

9.6 Questions and Answers

Question 9.1

Distinguish between an overlapping and a non-overlapping genetic code. What evidence shows that the code is non-overlapping?

Table 9.1 The genetic code

First letter	Second letter				Third letter
	U	C	A	G	
U	UUU ⎤ Phe UUC ⎦ UUA ⎤ Leu UUG ⎦	UCU ⎤ UCC ⎥ Ser UCA ⎥ UCG ⎦	UAU ⎤ Tyr UAC ⎦ UAA Term. UAG Term.	UGU ⎤ Cys UGC ⎦ UGA Term. UGG Trp	U C A G
C	CUU ⎤ CUC ⎥ Leu CUA ⎥ CUG ⎦	CCU ⎤ CCC ⎥ Pro CCA ⎥ CCG ⎦	CAU ⎤ His CAC ⎦ CAA ⎤ Gln CAG ⎦	CGU ⎤ CGC ⎥ Arg CGA ⎥ CGG ⎦	U C A G
A	AUU ⎤ AUC ⎥ Ile AUA ⎦ AUG Met	ACU ⎤ ACC ⎥ Thr ACA ⎥ ACG ⎦	AAU ⎤ Asn AAC ⎦ AAA ⎤ Lys AAG ⎦	AGU ⎤ Ser AGC ⎦ AGA ⎤ Arg AGG ⎦	U C A G
G	GUU ⎤ GUC ⎥ Val GUA ⎥ GUG ⎦	GCU ⎤ GCC ⎥ Ala GCA ⎥ GCG ⎦	GAU ⎤ Asp GAC ⎦ GAA ⎤ Glu GAG ⎦	GGU ⎤ GGC ⎥ Gly GGA ⎥ GGG ⎦	U C A G

The first letter of each of the 64 triplets is shown at the left, the second letter across the top and the third letter down the right-hand side. The three triplets designated Term. are the three chain-termination triplets. The table clearly shows how the degeneracy of the code is largely attributable to the relatively unimportant base in the third position.

Answer 9.1

Consider the nucleotide sequence

$$\text{C A G C U A G U A U C G}$$

If the code is fully overlapping, then the last two nucleotides of one codon will be the first two nucleotides of the succeeding codon and the successive codons will be

$$\text{CAG} \qquad \text{AGC} \qquad \text{GCU} \qquad \text{CUA} \dots \qquad \text{(A)}$$

or if the code is partially overlapping, when the last nucleotide of one codon is the first nucleotide of the succeeding codon, the successive codons will be

CAG   GCU   UAG   AGU ...   (B)

However, if the code is non-overlapping, the successive codons will be

CAG   CUA   GUA   UCG ...   (C)

If the code is overlapping, then there must be restrictions on the sequence of codons and on the sequence of amino acids in the polypeptide. Thus, if the code is fully overlapping, then the codeword CAG MUST be followed by one of the four codons beginning AGN and a particular amino acid can only be followed by one of four other amino acids. No such restrictions have ever been observed.

Furthermore, if the code is fully overlapping, then a single base pair substitution (mutation) in the DNA will affect THREE adjacent codons and, hence, THREE adjacent amino acids. Thus, a mutation which resulted in a G to A substitution at position 3 (above) would generate the mRNA

C A A C U A G U A U C G ...   (D)

and the successive codons would be

CAA   AAC   ACU   CUA ...   (E)

Compare this sequence with (A) above.

Similarly, if the code were only partially overlapping, then each mutation would affect two successive amino acids in the polypeptide. On the other hand, if the code is non-overlapping, then only ONE codon and ONE amino acid is altered; this is what is always observed.

Question 9.2

Outline the experiments that first established the general nature of the genetic code.

Answer 9.2

In 1961 Francis Crick, Sydney Brenner and their colleagues published the results of their far reaching experiments on the nature of the genetic code, using *rIIB* mutants of phage T4. The *rIIB* gene was particularly suitable for their studies, since the left-hand (promoter) end is relatively unimportant and wild type plaques are produced so long as the right-hand end of the gene is *read in phase* and *accurately translated*.

The plan of their experiments and the interpretation of their results is shown in Figure 9.1.

They started by isolating an *rIIB* mutant induced by the acridine dye proflavin; this slips in between adjacent base pairs in the DNA molecule and, when the DNA replicates, causes the addition or deletion of a single base pair. In these frameshift mutants the code is read out of phase beyond the site of the mutation; thus, the *rIIB* gene product has an incorrect amino acid sequence at its carboxy

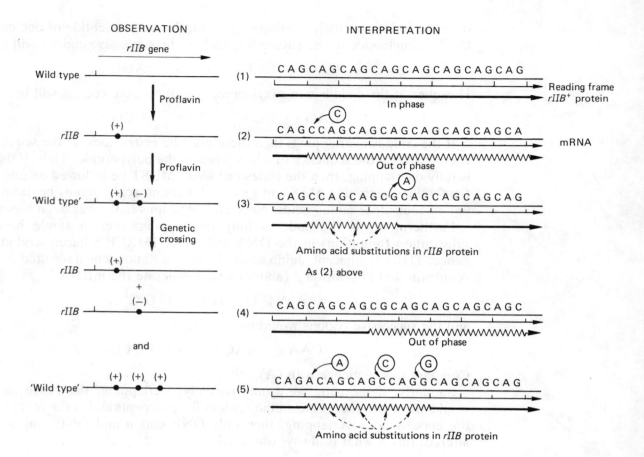

Figure 9.1 *Establishing the general nature of the genetic code. The interpretation at the right shows an arbitrary nucleotide sequence for the transcript of the* rIIB *gene (this is the same as the sequence along the antisense strand of the DNA duplex) and how the addition of a base pair (+) to the DNA adds a base to the mRNA, throwing the reading frame out of phase beyond the mutation. A second deletion mutation (−) restores the reading frame, with the result that only the codons between the two mutations are read out of phase (3). Likewise, three (+) or three (−) mutations result in localised misreading of the code and the addition or deletion of one codon from the message*

end and the result is an *rIIB* phenotype. The first mutant that mapped at the left-hand end of the *rIIB* gene they arbitrarily designated (+) (Figure 9.1.2).

They argued that treatment of this mutant with proflavin would induce further addition (+) or deletion (−) mutations and that, if a (−) mutation occurred close to the site of the original (+) mutation, the reading frame would be restored beyond the rightmost mutation; the code would be read out of phase between the two mutations and this would cause several amino acid substitutions but only in the relatively unimportant left-hand (amino) end of the *rIIB* protein (Figure 9.1.3), so that the double mutant would have a wild or nearly wild phenotype.

They found such wild type reversions and by making suitable crosses were able to separate the (+) and (−) mutations; as expected, these both turned out to have the *rIIB* phenotype (Figure 9.1.4).

They isolated many different (+) and (−) mutations within the left-hand end of the *rIIB* gene and found that:

(1) Most combinations of (+) and (−) mutations produced a wild phenotype.

(2) Double mutants containing two (+) or two (−) mutations were always mutant.

(3) When three (+) or three (−) mutations were recombined into a single strain, the triple mutant was always wild type — the consequence of these three mutations is to add or delete one complete codon from the non-essential end of the *rIIB* gene (Figure 9.1.5), so that the important right-hand end of the gene is read in phase.

(4) When the mutations in the triple mutants were of different signs, the strains were always mutant — the right-hand end of the gene was still read out of phase.

The last two observations are direct evidence for a triplet code (or, less likely, a code based on a multiple of three). The results also show that the code is commaless, as otherwise a single (+) or a single (−) mutation would only result in the misreading of a single codon — there would not be a reading frame to put out of phase and (+) and (−) mutations would be unable to compensate each other.

NOTE

The use of *rII* mutants in genetic crosses is described in Section 6.4.

Question 9.3

In their experiments with *rIIB* mutants of phage T4 Crick and Brenner recombined into a single genotype different pairs of base pair addition (+) and base pair deletion (−) mutations. Whereas most of the double mutants had a wild phenotype, others still retained the *rIIB* phenotype. Explain the latter observation.

Answer 9.3

(1) Addition or deletion of a base pair produces a frameshift mutation and all the downstream codons are read out of phase.

(2) In most instances the frameshift produces an in-phase chain termination triplet (UAG, UGA or UAA), causing the premature termination of translation:

WILD   CUA   GCU   AUG   AUG   UCA   GAU   CAU   . . .

rIIB mutant   CUA   CUA   UGA   UGU   CAG   AUC   AU. . . .   (−) mutation

⎯⎯⎯⎯⎯⎯⎯⎯⎯⎯| termination of translation

159

(3) If the (+) mutation occurs between the (−) mutation and the chain termination triplet, then the correct reading frame will be restored beyond the (+) mutation and the downstream end of the gene will be correctly translated

<u>CUA</u> <u>CUG</u> <u>AUG</u> <u>AUG</u> <u>UCA</u> <u>GAU</u> <u>CAU</u> . . .

(4) BUT if the (+) mutation is located downstream from the in-phase chain termination triplet, then, although the reading frame for the downstream end of the gene has been correctly restored, translation will still terminate at the in-phase chain termination triplet:

<u>CUA</u> <u>CUA</u> UGA UGU GCA <u>GAU</u> <u>CAU</u> . . .

———————————————‖ translation still terminates here

THUS, the mutant phenotype will always be expressed when there is an in-phase chain termination codon *between* the two frameshift mutations.

(Codons in the correct reading frame are underlined.)

Question 9.4

In an *in vitro* protein synthesising system poly(AC) was found to stimulate the incorporation of threonine and histidine, while poly(AAC) resulted in the production of polyasparagine, polythreonine and polyglutamic acid. Which of the codons present in these regular copolymers can be definitively assigned to particular amino acids?

Answer 9.4

Poly(AC) is ACACACACACACAC, which is read as

ACA   CAC   ACA   CAC   ACA ...

and these two codons must encode threonine and histidine.

Poly(AAC) is AACAACAACAACAACAAC ... and this can be read in three different ways (depending upon where reading commences), i.e.

as   AAC   AAC   AAC   AAC   AAC ...
or   ACA   ACA   ACA   ACA   ACA ...
or   CAA   CAA   CAA   CAA   CAA ...

encoding three different polypeptides, each made up of one repeating amino acid, polyasparagine, polythreonine and polyglutamic acid.

Comparing the two copolymers, there is only one common codon, ACA, and one common amino acid, threonine, incorporated into protein. Thus, ACA must be a codon for threonine.

Poly(AC) contains only the codons ACA and CAC. Since ACA is the codon for threonine, the remaining codon, CAC, must encode histidine.

AAC and CAA must encode asparagine and glutamic acid but from these data it is not possible to deduce which codon encodes which amino acid.

Question 9.5

In experiments using synthetic messengers Nirenberg found that polyUC (5U:1C) stimulated the incorporation of four amino acids into protein; these amino acids, and the relative frequencies with which they were incorporated, were:

phenylalanine   100,   serine 37,   proline 12,   leucine 5.

Given a copy of the genetic code (Table 9.1), calculate the expected relative frequencies of incorporation. Do these agree with the observed values?

Would you expect any other amino acids to have been incorporated in response to polyUC?

Answer 9.5

In a random 5U:1C copolymer the eight possible triplets occur at the following frequencies:

UUU	$5/6 \times 5/6 \times 5/6$	= 125/216	encodes phenylalanine
UUC	$5/6 \times 5/6 \times 1/6$	= 25/216	encodes phenylalanine
UCU	$5/6 \times 1/6 \times 5/6$	= 25/216	encodes serine
CUU	$1/6 \times 5/6 \times 5/6$	= 25/216	encodes leucine
CCU	$1/6 \times 1/6 \times 5/6$	= 5/216	encodes proline
CUC	$1/6 \times 5/6 \times 1/6$	= 5/216	encodes leucine
UCC	$5/6 \times 1/6 \times 1/6$	= 5/216	encodes serine
CCC	$1/6 \times 1/6 \times 1/6$	= 1/216	encodes proline

Thus, the expected codon frequencies in the message are:

phenylalanine	(125 + 25)/216	=	150/216
serine	(25 + 5)/216	=	30/216
proline	(5 + 1)/216	=	6/216
leucine	(25 + 5)/216	=	30/216

or 100:20:4:20, expressing the codon frequencies relative to UUU.

This is in general agreement with the observed ratio of 100:37:12:5, although much less leucine was incorporated than would have been expected.

NO other amino acids are expected to be incorporated into protein in response to polyUC, since all eight possible codons have been accounted for.

Question 9.6

Phage MS2 contains a molecule of single-stranded RNA which acts both as the phage chromosome and as messenger. The following is the coding sequence at the start of the coat protein gene and the corresponding amino acid sequence at the N-terminus of the coat protein:

codon	1	2	3	4	5	6	7	8	9	10	
nucleotide	AUG	GCU	UCU	AAC	UUU	ACU	CAG	UUC	GUU	CUC	...
amino acid		Ala –	Ser –	Asn –	Phe –	Thr –	Gln –	Phe –	Val –	Leu	...

(a) Why is there not a methionine residue at the N-terminus of the coat protein?

(b) What would be the effects on the amino acid composition of the coat protein if, as a result of mutation,
   i) an A is deleted from within codon 4,
   ii) the C is deleted from codon 4,
   iii) the U in codon 6 is replaced by a G,
   iv) the A in codon 6 is replaced by a G?

Answer 9.6

(a) The initiator AUG codon is recognised by a special tRNA charged with a formylated derivative of methionine (see Section 8.4). However, either the formyl group is enzymatically removed, leaving a terminal methionine, or, as here, the entire formylated methionine is cleaved from the completed protein.

(b) (i) If an A is deleted from codon 4, codon 4 and all the following codons will be read out of phase, thus:

AUG GCU UCU ACU UUA CUC AGU UCG UUC ...
Ala – Ser – Thr CTT

However, codon 5 is now read as UUA, which is a chain termination triplet — only a three amino acid long N-terminal fragment will be synthesised.

  (ii) The message is read as
AUG GCU UCU AAU UUA CUC AGU UCG UUC...
Ala – Ser – Asn CTT

A different three amino acid long N-terminal fragment is produced.

  (iii) This substitution affects only codon 6, which is changed from ACU to ACG. However, both ACU and ACG are codons for threonine, so this mutation has no effect on the coat protein.

  (iv) This substitution also affects only codon 6 and changes it from ACU to GCU. However, GCU codes for alanine so that the coat protein sequence is altered to

Ala–Ser–Asn–Phe–Ala–Gln–Phe–Val–Leu ...

Question 9.7

Many spontaneous mutations involve the substitution of one base pair for another in the DNA, and, when they occur within a coding sequence, may result in the replacement of one amino acid by another in the polypeptide. What other amino acids might replace a tryptophan residue as a result of a single base pair substitution in the DNA?

Answer 9.7

The *only* codon for tryptophan is UGG. There are three possible substitutions at each position of the codon, so that there are nine codons related to UGG by single base pair changes. These are:

(1) Changes in the first position (U)

AGG	arginine
CGG	arginine
GGG	glycine

(2) Changes in the second position

UAG	chain termination triplet
UCG	serine
UUG	leucine

(3) Changes in the third position

UGA	chain termination triplet
UGC	cysteine
UGU	cysteine

Thus, arginine, glycine, serine, leucine and cysteine are the only amino acids that can replace tryptophan as the result of single base pair substitutions in the DNA. The chain termination triplets generated within the gene cause the premature termination of translation and result in the production of N-terminal polypeptide fragments.

9.7 Supplementary Questions

9.1 How many codons could be provided by a four-letter quadruplet code?

9.2 Explain what is meant by *reading frame*.

9.3 Why do most frameshift mutations result in the production of prematurely terminated polypeptides?

9.4 In an *in vitro* system polyU stimulates the incorporation of phenylalanine into protein, while polyA stimulates the incorporation of lysine. What amino acids would be incorporated in response to a mixture of polyA and polyU?

9.5 What polypeptides would be synthesised in an *in vitro* system in response to (**a**) poly(AUG) and (**b**) poly(UGAC)?

9.6 What are the relative frequencies of the eight triplets containing only uracil and cytosine in a random polynucleotide containing uracil, cytosine and guanine in the ratio 2U:1C:1G?

9.7 A protein produced by a wild type strain of *E. coli* has a valine residue at position 108. Three mutants (A, B and C) are isolated and found to have glycine, glutamic acid and leucine substitutions at this position. Reversions of mutants A and B had, respectively, tryptophan and lysine residues at position 108; mutant C was not tested further. Explain these results and deduce the codons present in each of six strains.

9.8 Which crosses between strains A, B and C (Question 9.7) would produce wild type recombinants as a result of crossing-over within the codon at position 108?

9.9 A wild type protein has a methionine residue at a particular position. List the possible amino acid substitutions that might occur as the result of a single base pair substitution in the corresponding DNA (you may use the table of the genetic code, Table 9.1).

9.10 In an *in vitro* system polyUG stimulates the incorporation into protein of phenylalanine and cystine but not of alanine. In 1962 Chapeville treated charged cysteine tRNA with Raney nickel, so reducing the attached cystine to alanine; when this altered amino acyl tRNA was used in the *in vitro* system, alanine was incorporated into protein. What is the significance of this result?

10 Genetic regulation in procaryotes

10.1 Inducible and Repressible Systems

Since the early days of biochemical genetics it has been clear that cells must be able to regulate their genetic activity. This is particularly clear in higher organisms, where there is very good reason to suppose that almost every cell contains a complete set of genetic information, and it would be very wasteful if, for example, the secretory cells of the pancreas, whose main function is to produce digestive enzymes, were to produce the structural protein keratin, normally found in nails, horns and hooves. In micro-organisms it was known as early as 1900 that yeast cells only produced the enzymes enabling them to grow on the sugar lactose when lactose is actually present and that, if the cells are then transferred to a non-lactose-containing medium, these enzymes are 'lost': in other words, the presence of the substrate (lactose) specifically **induces** the production of these enzymes. These enzymes are said to be **inducible** and the substrate that induces their production is the **inducer**. In contrast, the **repressible** systems were only discovered in 1953 by Jacques Monod. He found that in the presence of tryptophan, the end product of a biochemical pathway, cells of *E. coli* no longer produced the enzyme tryptophan synthetase which is required for the biosynthesis of tryptophan. Thus, in a repressible system the enzymes required for the synthesis of an end product are not synthesised in the presence of that end product.

Inducible systems are characteristic of **catabolic pathways**, whereas repressible systems are generally those involved in **anabolism**.

10.2 The Clustering of Genes

In bacteria the genes involved in a particular biosynthetic pathway tend to be arranged in tight clusters on the bacterial chromosome and to be transcribed onto a single species of mRNA. Thus, in *E.coli* the five genes involved in tryptophan biosynthesis not only are tightly clustered on the chromosome, but also are arranged in the same order as the sequence of the reactions that their products catalyse (Figure 10.1).

This clustering of related genes was first observed by Miloslav Demerec in 1956 but its significance remained obscure until 1961, when Jacques Monod and Francois Jacob proposed their **operon model** for the genetic control of protein synthesis. They realised that genetic regulation occurs primarily at the level of transcription and, because these clusters of genes of related function are transcribed onto a single molecule of messenger RNA, their expression could be co-ordinately controlled by permitting or preventing transcription; this type of regulation is possible because bacterial messengers are rapidly degraded and in the absence of mRNA the genes cannot be translated into protein. They called

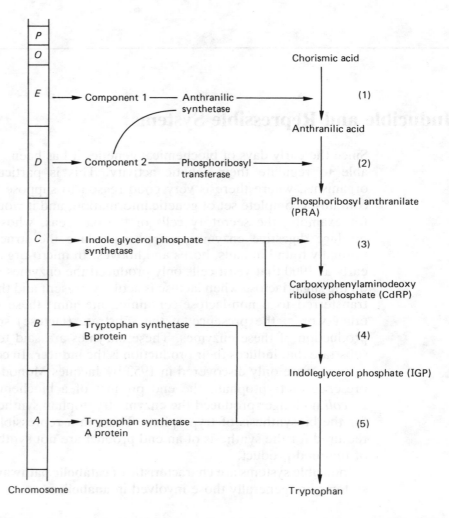

Figure 10.1 *The biosynthesis of tryptophan in E. coli. Tryptophan is synthesised from chorismic acid by five sequential reactions. The trpE protein (ASase component 1) is only active when complexed with the trpD gene product (ASase component 2) to form anthranilic synthetase. Reaction 1 is catalysed by this complex enzyme but reaction 2 is catalysed by ASase component 2 (otherwise known as PR transferase) alone. Reactions 3 and 4 are catalysed by IGP synthetase, the trpC gene product, while reaction 5 requires tryptophan synthetase, a complex of the trpA and trpB gene products*

these groups of co-ordinately regulated genes **operons**, and their operon model explains how this regulation is achieved.

10.3 The *lac* Operon of *E. coli* Inducible

The *lac* operon (Figure 10.2), intensively studied by Monod and Jacob, consists of three structural genes and three genetic control sequences:

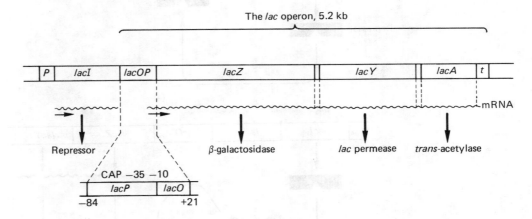

Figure 10.2 *The lactose operon of* E. coli. *The* lacZ–lacY *and* lacY–lacA *genes are separated by short untranslated sequences*

Structural genes:

lacZ encoding **β-galactosidase**, the enzyme which hydrolyses lactose to galactose and glucose;

lacY encoding a **permease** which transports lactose across the cell membrane, thus enabling it to enter the cell;

lacA encoding a **transacetylase;** its function is not understood and it is not essential for lactose catabolism.

Genetic control sequences:

lacP a **promoter** sequence, some 100 bp long, to which RNA polymerase binds and at which it initiates transcription (Section 8.2);

lacO an **operator** sequence, some 45 bp long, overlapping the promoter and to which the *lacI* repressor can bind;

lac-t a **terminator** sequence signalling the termination of transcription (Section 8.2).

Note that both *lacP* and *lacO* are discrete units of genetic function, and, although they overlap, each can properly be regarded as a gene; the complex control region including these **regulatory elements** is referred to as the *lacOP* region.

These components constitute the *lac* **operon** and expression of the three structural genes within it is controlled by the protein product of the *lacI* gene, the *lac* **repressor**. In the absence of lactose, this repressor can recognise and bind to the specific nucleotide sequence of the operator; this prevents the initiation of transcription by preventing RNA polymerase from binding to the adjacent promoter sequence, and the operon is said to be **repressed** (Figure 10.3a).

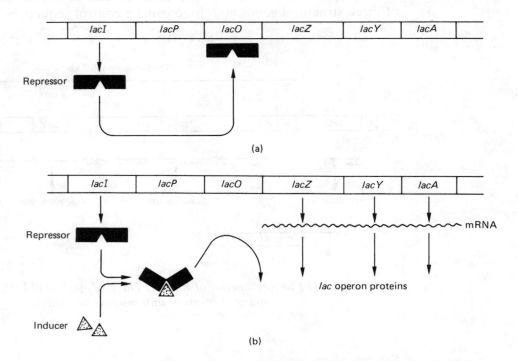

Figure 10.3 *Induction of the* lac *operon.*
(a) In the absence of inducer molecules, the lacI *repressor binds to the operator and switches off the operon.*
(b) When substrate (inducer) is present, it binds to and inactivates the repressor; the repressor complex cannot bind to the operator and the structural genes are transcribed and translated

Even when the operon is repressed, there is still a very low level of *lac* operon transcription, and repressed cells contain between 0.5 and 5 molecules of β-galactosidase per cell.

Repression occurs provided that the substrate (lactose) is not present, but whenever lactose is present, the operon is transcribed and is said to be **derepressed** or **induced**. In many inducible operons the inducer is the substrate itself, but in the *lac* operon the inducer is **allolactose**, a by-product of lactose catabolism produced by the activity of the few molecules of β-galactosidase present in the repressed cells. These molecules of allolactose bind to the repressor, causing it to undergo a conformational change. In this altered form the repressor is no longer active and cannot recognise the operator sequence; thus, RNA polymerase can bind to the promoter and the operon can be transcribed (Figure 10.3b). After induction there are between 1000 and 10 000 molecules per cell of β-galactosidase.

Hence, the substrate (or, more specifically, allolactose) acts as an **inducer** by combining with and inactivating the repressor, and the *lac* operon mRNA and proteins are only produced when actually required by the cell.

This type of control is referred to as **negative control**, since the operon is only switched on in the **absence** of the active repressor protein.

Important points to note are:

(1) the repressor gene (*lacI*) is *not* part of the operon.

(2) The repressor is a diffusible protein and there is no requirement for the repressor gene to be located adjacent to the operon it controls.

(3) Repressor prevents transcription by binding to the operator and hindering the binding of RNA polymerase to the promoter; thus, the operator must be immediately adjacent to the promoter.

(a) Mutations Affecting Expression of the *lac* Operon

Mutants with altered *lac* operon expression result from mutations either within the structural genes of the operon or within the regulatory sequences, and they can be distinguished by (1) their effect on the phenotype and (2) their behaviour in F′ partial diploids.

Mutations in the lac *structural genes*   A missense mutation within a structural gene may cause a single amino acid substitution in the peptide encoded by that gene, and usually only affects the expression of that one gene. For example, a mutation in *lacZ* may result in the absence of β-galactosidase but the *lacY* permease and the *lacA trans*-acetylase will continue to be normally inducible; similarly, a *lacY* mutant will lack the permease and continue to produce normal amounts of β-galactosidase and the *trans*-acetylase. Note that the activity of the unmutated genes remains inducible.

Mutations in the lac *regulatory elements*   In contrast to mutations within the structural genes, mutation either within *lacI* or within the *lacO* and *lacP* control sequences will co-ordinately affect expression of all three *lac* operon structural genes; this is how these elements were first detected.

lacI⁻   The repressor protein is either missing or inactive; the *lac* structural genes are permanently switched on and synthesis is **constitutive**.

*lacI*ˢ   The repressor protein is altered ('super-repressor'), so that it is unable to bind to inducer molecules; thus, the repressor remains permanently bound to the operator and the operon is permanently switched off.

*lacO*ᶜ   The sequence within the operator has been altered so that it is no longer recognised by the repressor; thus, the three structural genes are constitutively expressed (O^c, operator-constitutive)

lacP⁻   The promoter sequence is altered and has a reduced capacity to bind RNA polymerase and to initiate transcription; reduced amounts of the *lac* operon mRNA and, hence, of the three gene products are present.

The phenotypes of these mutant strains are shown in Table 10.1.

Table 10.1 Expression of the *lacZ* and *lacY* genes

	Inducer present		Inducer absent	
	β-galactosidase	permease	β-galactosidase	permease
$I^+O^+Z^+Y^+$	+	+	−	−
$I^+O^+Z^-Y^+$	−	+	−	−
$I^+O^+Z^+Y^-$	+	−	−	−
$I^-O^+Z^+Y^+$	++	++	++	++
$I^s\,O^+Z^+Y^+$	−	−	−	−
$I^+O^c\,Z^+Y^+$	++	++	++	++

(b) Catabolite Repression

Glucose is normally the preferred energy source and, when abundant, it would be wasteful for the cell to make the enzymes involved in lactose catabolism; under these conditions not only the *lac* operon but also the other inducible catabolic operons (e.g. the arabinose and galactose operons) are co-ordinately switched off. This is glucose inhibition, and the effect is called **catabolite repression**. Thus, these operons are subject to a global control which only permits them to be switched on (1) if glucose is absent and (2) the appropriate substrates (lactose, arabinose, galactose, etc.) are present.

A typical promoter (Chapter 8) includes two special sequences, one centred on -35 (i.e. 35 nucleotides upstream from the transcriptional start), where RNA polymerase initially binds, and the Pribnow box centred on -10, where transcription is initiated. In the operons sensitive to glucose inhibition there is a third sequence, the **CAP binding site**, which, in the *lac* operon, is centred around -61, -62.

In the absence of glucose, cyclic AMP is abundant and forms a complex with CAP, the catabolite activator protein. This cAMP–CAP complex binds to the CAP site within the promoter and, in an unknown way, stimulates the binding of RNA polymerase to the -35 region. In the presence of glucose, cAMP levels are greatly reduced, little or no cAMP–CAP complex is present and the promoter is not activated. Thus, the cAMP–CAP complex is a **positive regulator** and transcription can *only* occur when this complex is bound to be promoter.

The *lac* operon and the other glucose-sensitive operons are, therefore, co-ordinately regulated by the levels of cAMP in the cell, and this, in turn, depends upon the amount of glucose present.

10.4 The Tryptophan Operon

(a) Repression

This is a good example of a **repressible operon**. In repressible operons the structural genes are not transcribed in the presence of the end product (tryptophan). In the *trp* operon (Figure 10.4) the *trpR* regulatory gene is *not* adjacent to the *trp* operon and its immediate product, the *trpR* repressor, is

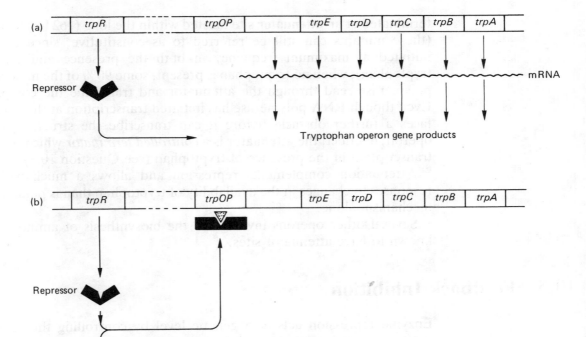

Figure 10.4 *Repression of the* trp *operon.*
(a) In the absence of co-repressor (tryptophan), the trpR *repressor cannot bind to the operator and the structural genes are transcribed and translated.*
(b) When co-repressor is present, it binds to the trpR *repressor; this can now bind to the operator and the operon is switched off*

inactive and is unable to recognise and to bind to the operator sequence; thus, the operon is switched on. However, when tryptophan is present, it combines with the repressor to produce an active repressor complex which binds to the operator and switches off the structural genes. Thus, in a repressible system an *inactive repressor* is *activated* by combining with the *end product* of the pathway; this end product is sometimes referred to as **co-repressor**.

Neither the *lac* nor the *trp* repressors are subject to genetic regulation; both *lacI* and *trpR* are transcribed all the time, irrespective of the presence of the substrates or end products of the operons — they are synthesised constitutively.

(b) Attenuation

The *trpR* repressor exerts a coarse control over expression of the *trp* structural genes by preventing transcription in the presence of high levels of tryptophan, but there is a further mechanism, known as **attenuation**, which permits a fine level of control when the operon is derepressed. This effect is most clearly seen in derepressed mutants ($trpR^-$, $trpO^c$) showing constitutive *trp* expression. When tryptophan is abundant, there is only 10% transcription of the complete operon, as 90% of the transcripts initiated at the promoter are prematurely terminated at an **attenuator** site located within the long (162 bp) leader sequence (these mutants can still be referred to as constitutive, since transcription is initiated at maximum frequency in both the presence and the absence of tryptophan); when no tryptophan is present, some 90% of the molecules of RNA polymerase read through the attenuator and transcribe the complete operon. Even though RNA polymerase has initiated transcription at the *trp* promoter, it faces a further obstacle before it can transcribe the structural genes in the operon; in effect, the attenuator is a *controlled terminator* which only terminates transcription in the presence of tryptophan (see Question 10.8).

Attenuation complements repression and allows a much wider range of responses to changes in the availability of tryptophan than is possible with either mechanism alone.

Several other operons involved in the biosynthesis of amino acids are also known to have attenuator sites.

10.5 Feedback Inhibition

Enzyme repression acts at a genetic level by controlling the rate of enzyme synthesis and, hence, the **amount** of enzyme present in the cell, but in many biochemical pathways leading to the production of small essential metabolites, such as the amino acids, there is a further mechanism of control acting purely at a biochemical level; this is **feedback inhibition**, an efficient method for achieving fine control by regulating enzyme **activity**.

If *E. coli* is grown in minimal medium, the *trp* operon is derepressed, the *trp* enzymes are synthesised and tryptophan is produced, but if excess tryptophan is suddenly added, then repression is re-established and synthesis of the *trp* enzymes ceases. However, in these newly repressed cells the *trp* enzymes are still present and yet the synthesis of tryptophan ceases immediately; this is because feedback inhibition prevents the wasteful synthesis of tryptophan.

In the usual type of feedback inhibition the activity of the first enzyme in the pathway is inactivated by binding to molecules of the end product of the pathway. Thus, the enzyme has two binding sites, one for binding the substrate molecules and the other binding the end product. When excess end product is present, molecules bind to the enzyme, causing it to undergo a conformational change, so restricting its activity. In the tryptophan pathway feedback inhibition occurs by molecules of tryptophan binding to anthranilic synthetase, thus reducing its ability to bind chorismic acid and glutamine (Figure 10.5).

This inhibition is usually reversible, so that there is a gradual resumption of enzyme activity as the concentration of the end product decreases.

Proteins which have two separate but interacting binding sites are referred to as **allosteric proteins**.

10.6 Control at the Level of Translation

From the point of view of economy, it is better for a cell to produce mRNA only when the gene products it encodes are required than it is to produce the mRNA and to regulate the initiation of translation. Nevertheless, there are instances where regulation occurs at the level of translation.

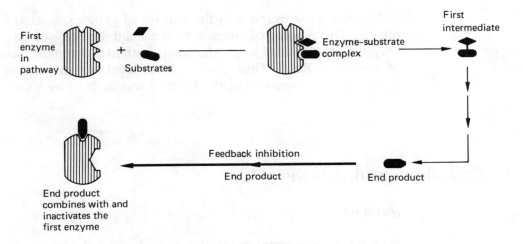

Figure 10.5 *Feedback inhibition. The first enzyme in the pathway has two separate and interacting binding sites. The end product of the pathway binds to one of these sites; the enzyme undergoes a conformational change and loses its catalytic activity*

Synthesis of the ribosomal proteins is a good example of translational control. Each ribosome contains about 50 proteins and the genes encoding these proteins are organised into at least 20 operons; these operons can only be regulated inefficiently at the level of transcription and the primary control is at the level of translation. One of the simplest operons is the L11 operon (Figure 10.6).

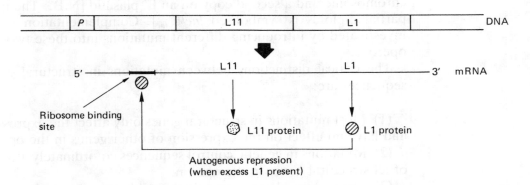

Figure 10.6 *Translational repression of the L11 ribosomal protein operon in* E. coli. *The L1 ribosomal protein can bind to the 5′ end of the mRNA for the L11 operon and block the ribosome binding site*

Normally, as the ribosomal proteins are synthesised they complex with the ribosomal RNAs and are assembled into ribosomes, but if these proteins are in excess, as when enough ribosomes have been assembled, then translation of the ribosomal proteins ceases. In the L11 operon this is achieved by the L1 protein, which, when in excess, binds to the 5′ end of the mRNA near the ribosomal binding site and prevents translation of all the structural genes in the operon; this is **autogenous repression**. Thus, the synthesis of the L1 and L11 proteins is co-ordinated with the process of ribosome assembly.

173

Another good example is the control of gene expression in the small RNA phages. The genome of phage MS2 is a small molecule of single-stranded RNA, only 3569 nucleotides long, which acts both as a chromosome and as a molecule of messenger RNA. Thus, control at the level of transcription is impossible and gene regulation must be at the level of translation (see Question 10.9).

10.7 Questions and Answers

Question 10.1

How can simple complementation tests distinguish between mutations in the *lac* operon structural genes and in their genetic control sequences?

Answer 10.1

In their first experiments Jacob and Monod (1961) carried out complementation tests, using the newly discovered F′ plasmids (see Section 6.3d). An F′ plasmid carrying a copy of the *lac* operon was transferred into an F⁻ recipient strain; this created a stable partial diploid with one copy of the *lac* operon on the *E. coli* chromosome and a second copy on an F′ plasmid [N.B.: The genotype of these partial diploids is symbolised *lac*/F′$_{lac}$]. Complementation relationships were investigated by introducing different mutations into these two copies of the *lac* operon.

The critical distinctions between mutations in structural genes and control sequences are:

(1) Most mutations in structural genes only affect the expression of that gene and have no effect on the expression of other genes in the operon.

(2) Mutations in genetic control sequences co-ordinately affect the expression of all structural genes in the operon.

(3) The structural genes encode diffusible gene products, so that both *lacZ*+ *lacY*−/F′$_{lacZ-~lacY+}$ and *lacZ*+ *lacY*+/F′$_{lacZ-~lacY-}$ partial diploids are phenotypically *lac*+. The defect in one gene is made good by the corresponding wild type gene on the other genetic element, and vice versa. Hence, the wild type alleles are **dominant in both the *cis* and *trans* arrangements**.

(4) Genetic control sequences only control the expression of those structural genes with which they are associated on the same genetic element, i.e. in the *cis* arrangement. Thus, *lacP*+ *lacZ*+/F′$_{lacP-~lacZ-}$ partial diploids are wild type, whereas *lacP*+ *lacZ*−/F′$_{lacP-~lacZ+}$ are unable to ferment lactose. Transcription can only be effectively initiated from a wild type promoter, so that β-galactosidase can only be made when both *lacP*+ and *lacZ*+ are on the same genetic element.

That is, mutations in regulatory sequences are ***cis*-dominant and *trans*-recessive.**

Question 10.2

What are the differences between a *lacI*⁻ and a *lacZ*⁻ mutation? How would you show that *lacI*⁺ encodes a diffusible regulatory protein?

Answer 10.2

(1) (i) P1-mediated transduction experiments would establish within which gene any newly isolated mutation was located. All *lacI* mutations map to the left of all *lacZ* mutations.

(ii) *lacZ*⁻ mutations abolish β-galactosidase activity but the permease and transacetylase activities are normally inducible. In *lacI*⁻ mutants all the *lac* operon enzymes are synthesised constitutively. Thus, *lacZ*⁻ mutants cannot ferment lactose, whereas *lacI*⁻ mutants can.

(2) Since *lacI*⁻ co-ordinately affects the expression of all the structural genes in the *lac* operon, it must either (a) encode a regulatory protein (inducer or repressor) or (b) be a genetic control sequence.

If (a) is correct, then *lacI*⁺ will be dominant to lacI⁻ in a partial diploid, whereas if (b) is correct, *lacI*⁺ will be *cis*-dominant and *trans*-recessive. Therefore, construct the following partial diploids and examine their phenotypes:

lacI⁻ *lacZ*⁺/F'$_{lacI+ lacZ-}$: If (a) is correct, this strain will be inducible and able to grow on minimal medium + lactose; if (b) is correct, the strain will not produce β-galactosidase and will be unable to ferment lactose.

lacI lacZ⁻/F'$_{lacI+ lacZ+}$: On either hypothesis the *lac* operon is inducible. This *trans* arrangement confirms that *lacI*⁻ is *cis*-dominant and *trans*-recessive.

NOTE

Remember that genetic control sequences only affect the expression of genes with which they are associated in the *cis* arrangement.

Question 10.3

What are the phenotypes of the following partially diploid strains of *E. coli*? Briefly explain your answers.

(a) *lacI*⁺ *lacO*^{c} *lacZ*⁺/F'$_{lacI- lacO+ lacZ-}$

(b) *lacI*⁻ *lacO*^{c} *lacZ*⁺/F'$_{lacI+ lacO+ lacZ-}$

(c) *lacI*⁺ *lacO*⁺ *lacZ*⁺/F'$_{lacI- lacO^c lacZ-}$

(d) *lac I*^{s} *lacO*^{c} *lacZ*⁺/F'$_{lacI- lacO+ lacZ-}$

(e) $lacI^s$ $lacO+$ $lacZ+$/$F'_{lacI^-\ lacO^c\ lacZ^-}$

(f) $lacI+$ $lacO+$ $lacP-$ $lacZ+$/$F'_{lacI^s\ lacO^c\ lacP^+\ lacZ^+}$

Answer 10.3

(a) β-galactosidase is synthesised constitutively. Although $lacI+$ repressor is present, it cannot bind to the $lacO^c$ operator and $lacZ+$ is expressed constitutively.

(b) β-galactosidase is produced constitutively. Since $lacI+$ encodes a diffusible regulatory protein, it is irrelevant whether it is in the *cis* or the *trans* arrangement.

(c) β-galactosidase production is inducible. In this partial diploid $lacZ+$ is associated in the *cis* with a wild type operator sequence; since repressor is present, $lacZ+$ expression is inducible. $lacZ-$ is transcribed constitutively but this has no effect on the phenotype.

(d) Super-repressor is present and binds to the $lacO+$ sequence but not to $lacO^c$; thus, β-galactosidase is constitutively expressed.

(e) This is like **(d)** except that $lacO+$ and $lacZ+$ are in the *cis* arrangement. Thus, β-galactosidase synthesis is permanently switched off. It is of no consequence that the $lacZ-$ gene is transcribed constitutively.

(f) Super-repressor is present but the operon on the chromosome is switched off because the promoter is mutant. The $lacO^c$ mutation on the F′ operon prevents the binding of either repressor or super-repressor, so that this $lacZ+$ gene is transcribed constitutively.

Question 10.4

E. coli is grown in liquid minimal medium containing a neutral carbon source (e.g. glycerol) which does not induce the *lac* operon. After 1 h lactose is added to the growth medium and after a further period an excess of glucose is added. How do these changes affect the expression of the *lac* operon?

Answer 10.4

For the first hour no inducer (lactose) is present, so that the *lac* operon is switched off. When lactose is added, after 1 h, then, since glucose is absent, the *lac* operon is induced and lactose is fermented. After a further period, the addition of a large amount of glucose results in catabolite repression and the *lac* operon is again switched off.

Question 10.5

Explain what is meant by a *gratuitous inducer*.

Answer 10.5

This is a compound structurally related to the natural inducer, and although it can induce the operon, it is not metabolised by any of its gene products. In the *lac* operon isopropylthiogalactoside (IPTG) is an analogue of lactose and can replace allolactose as an inducer, but it is not metabolised by β-galactosidase.

Question 10.6

When a *lacZ⁻* or a *lacY⁻* mutant is grown in the presence of lactose, the remaining genes in the *lac* operon are not induced. Explain why this is so.

Answer 10.6

Allolactose is the natural inducer of the lactose operon and this is metabolised from lactose by the few molecules of β-galactosidase present in uninduced cells.

In *lacZ⁻* mutants β-galactosidase is totally absent and lactose cannot be metabolised; in *lacY⁻* mutants the permease is totally absent and lactose is unable to enter the cell. In neither mutant can allolactose be produced, and so the remaining genes in the operon cannot be induced by lactose present in the medium.

However, the remaining genes can be induced by a gratuitous inducer such as IPTG — this acts as an inducer and can enter the cell without the assistance of the *lacY* permease.

Question 10.7

Distinguish between inducible and repressible gene control.

Answer 10.7

In an inducible system the operon is switched on only in the presence of inducer. In the absence of inducer, repressor binds to the operator and prevents transcription of the structural genes; when inducer is present, it binds to the repressor and alters its configuration so that it can no longer bind to the operator — thus, the operon is switched on. Enzyme induction is characteristic of catabolic pathways, and the inducer is a molecule of substrate or a closely related molecule.

In a repressible system the operon is turned off in the presence of the end product. In the absence of end product, the repressor cannot bind to the operator and the operon is switched on; when end product is present, it binds to the repressor and alters its configuration, so that it can now bind to the operator — thus, the operon is switched off. Enzyme repression is characteristic of anabolic pathways.

Question 10.8

Explain how attenuation is thought to modulate expression of the *E. coli* tryptophan operon.

Answer 10.8

Attenuation modulates expression of the derepressed *trp* operon according to the availability of charged tRNA^{Trp} (tryptophanyl-tRNA), which, in turn, is dependent upon the level of tryptophan in the cell.

The leader sequence of the *trp* mRNA is unusually long and includes all the genetic information necessary to encode a polypeptide 14 amino acids long (including an AUG start and a UGA chain termination codon); this peptide contains two adjacent Trp residues, so that tryptophanyl-tRNA is necessary for translation of the leader peptide. The essential requirements for regulation by attentuation are (1) the translation of this leader peptide and (2) the coupling of transcription and translation, so that as RNA polymerase is transcribing the leader, so ribosomes are following closely behind along the newly transcribed messenger and translating the leader peptide.

The mRNA leader sequence includes two pairs of similar inverted repeats (designated 1, 2, 3 and 4 in Figure 10.7); sequence 2 is partially base-complementary to both sequences 1 and 3, so that either 1 + 2 or 2 + 3 or 3 + 4 or 1 + 2 and 3 + 4 can base pair and form stem-and-loop structures.

The stem-and-loop structure formed by the pairing of sequences 3 and 4 is nearly identical with the *trp* operon terminator (Figure 8.11) and, like the terminator, includes a tail of seven consecutive Us on the 3′ side of the stem. Whenever this attenuator structure forms, it acts as a terminator and transcription ceases.

Note that the two Trp codons fall within sequence 1 and that the chain termination codon for the leader peptide is between sequences 1 and 2. Thus, if tryptophan is abundant, then charged tRNA^{Trp} is plentiful and the ribosomes will translate the leader sequence as far as the UGA stop codon. The bound ribosomes mask both regions 1 and 2, so that the stem and loop structures 1 + 2 and 2 + 3 cannot form; this allows the 3 + 4 attenuator loop to form and to act as a terminator, causing the molecule of RNA polymerase to dissociate from the DNA template.

On the other hand, if tryptophan is scarce, then very little tryptophanyl-tRNA will be available and the ribosomes will stall at the two Trp codons. This masks region 1 (only) and allows the stem-and-loop structure 2 + 3 to form before sequence 4 has been transcribed. The formation of this structure prevents formation of the 3 + 4 structure and termination is prevented. Thus, read-through occurs and the rest of the *trp* operon is transcribed.

In effect, it is the position of the ribosomes on the mRNA which determines the secondary structure of the mRNA and whether attenuation occurs.

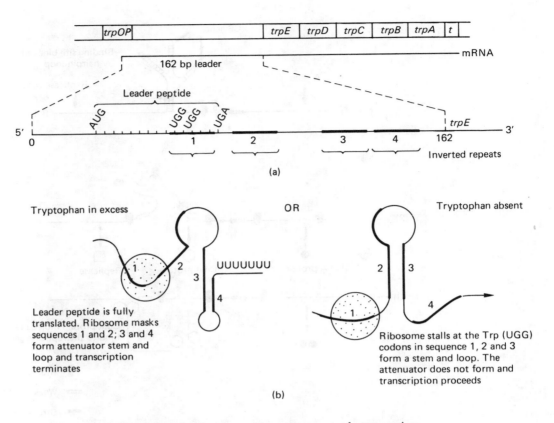

Figure 10.7 *The tryptophan operon and attenuation.*
(a) The tryptophan operon and the structure of the leader sequence.
(b) Attenuation occurs when excess tryptophan is present and the leader sequence is actively translated

Question 10.9

Describe the control of translation in the RNA phage MS2.

Answer 10.9

The RNA genome of MS2, like many RNA molecules, has a complex secondary structure consisting of many hairpin loops held together by H bonding between pairs of complementary bases. The genome consists of only four genes and, following infection of an *E. coli* host, the three principle genes (Figure 10.8) are differentially translated. Not only are more molecules of coat protein made than of either the A protein (required for encapsidation) or the replicase (required for RNA replication), but also the replicase gene is not translated until *after* translation of the CP gene has commenced, and even then this translation ceases about 10 min after infection. The basis of this control is as follows:

(1) Initially the A and the CP genes are independently translated. More molecules of the coat protein are made, because this gene has the stronger

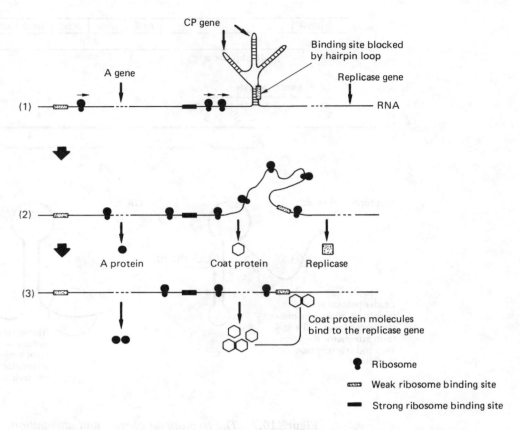

Figure 10.8 *Translational control of gene expression in phage MS2.*
(1) Ribosomes translate the A gene (weakly) and the coat protein gene (strongly).
(2) Ribosomes translating the CP gene open up the hairpin, freeing the binding site of the replicase gene.
(3) Translation of the replicase gene is switched off

ribosome binding site and so more ribosomes will pass along the gene and translate it.

(2) The replicase gene is only translated after the CP gene has been translated. This is because the ribosome binding site and AUG start codon for the replicase gene lie within a hairpin loop and so are not immediately available for ribosome binding. This hairpin loop involves sequences at the start of the CP gene, and as the ribosomes move along and translate the CP gene, so they will open up this hairpin loop. This enables the ribosomes to bind and to translate the replicase gene (Figure 10.8.2).

(3) The binding site for the replicase gene is weak and only a few molecules of replicase are made. Even so, after 10 min translation of this gene ceases. This is because molecules of the coat protein, when they reach a certain concentration, bind to the replicase gene and prevent any further translation.

10.8 Supplementary Questions

10.1 Distinguish between positive and negative control.

10.2 In most bacterial operons the structural genes are all closely linked and regulated from a single operator–promoter region. In some instances the structural genes are dispersed around the chromosome. How can the co-ordinated regulation of these genes be most simply achieved?

10.3 Distinguish between (a) up-promoter and down-promoter mutations and (b) upstream and downstream sequences.

10.4 Explain why operators and promoters are *cis*-dominant and *trans*-recessive, whereas genes encoding repressor proteins are both *cis*- and *trans*-dominant.

10.5 List some types of mutants that (a) are unable to synthesise a particular polypeptide and (b) synthesise the polypeptide constitutively.

10.6 What are the phenotypes of the following strains:
  (a) $trpA^-/F'_{trpP^-\ trpB^-}$,
  (b) $trpO^c/F'_{trpB^-}$,
  (c) $trpR^-\ trpA^-/F'_{trpO^c\ trpB^-}$?

10.7 In the *trp* operon of *E. coli* deletions at the *trpE* end of the operon result in the absence of all the *trp* gene products. On the other hand, deletions at the *trpA* end of the operon produce strains where *trpE*, *trpD*, *trpC* and *trpB* remain inducible. The five genes are closely linked and in the order *trpEDCBA*. What can you deduce regarding the location of the regulatory sequences and the direction of transcription?

10.8 Melibiose is a weak inducer of the *lac* operon and it is normally transported into the cells by its own permease, but if the cells are grown at 42 °C, this permease is inactivated and melibiose can only enter the cell if the *lacY* permease is present; thus, *lacY*⁻ and *lacP*⁻ mutants cannot grow at 42 °C on medium containing melibiose as the sole carbon source.
  How can these properties be used to isolate constitutive mutations within the *lac* operon?

10.9 Why are lysogenic cells immune to superinfection by the same or a very closely related phage? What type of phage mutants do you suppose would be able to grow on lysogenic cells?

10.10 How does glucose influence the expression of several different operons (the glucose-sensitive operons), all involved in the utilisation of sugars?

11 Mutation

11.1 Types of Mutation

Mutation is any heritable change in the nucleotide sequence of a genome; the process of mutation is **mutagenesis** and an organism showing an altered phenotype as the result of a mutation is a **mutant**. Mutation is the ultimate source of all new genetic variation and so has played a vital role in the processes of adaptation and evolution; mutants are the working tools of the geneticist, and without mutants none of the studies described in this book would have been possible.

Mutation, in its widest sense, includes such diverse phenomena as changes in chromosome number (polyploidy) and changes in the gross structure of eucaryotic chromosomes. Although these **chromosome aberrations** are of considerable interest to evolutionary geneticists and plant breeders, they are of comparatively little importance to the molecular geneticist; our attention will focus on **gene** or **point mutation** — changes occurring primarily within the genes themselves.

Not all mutations produce recognisable mutants. Some mutations will, for example, occur in non-coding regions of the genome and have no detectable effect on the phenotype. These **silent** mutations are probably frequent in higher eucaryotes, where the larger part of the genome is non-coding.

In higher organisms mutations can occur at any stage of development. Mutations occurring during somatic development may give rise to patches of mutant tissue (the size of the patch depending on the time during development at which mutation occurred) but these **somatic mutations** do not affect the gametes and cannot be passed on to the next generation. However, **germ-line** or **germinal mutations** occur in the sex cells, mutant gametes are produced and the mutation can be transmitted; all the mutations in higher eucaryotes described in this book are germinal mutations.

11.2 Some Types of Mutant Used in Genetic Studies

Mutations can be somewhat arbitrarily classified according to the most conspicuous effect they produce. The types of mutant include:

(1) **Visible or morphological mutants** The mutation directly affects the appearance of an organism, e.g. vestigial wing in *Drosophila*, albino mammals, plaque morphology mutants of phages.

(2) **Behavioural mutants** The mutation affects the behavioural pattern of the organism. In *Drosophila* 'dunce' is defective in its ability to learn, while 'amnesiac' learns well but soon forgets.

(3) **Sex-linked recessive lethals** These are mutations on the X chromosomes of higher eucaryotes which are lethal in homozygous females and hemizygous males; they have been extensively used for assessing the effect on mutation rate of different mutagenic treatments (Question 11.9).

(4) **Nutritional or biochemical mutants** These lead to the loss of a specific biochemical function. They include auxotrophs in micro-organisms (Section 7.4) and many human biochemical diseases, such as phenylketonuria (Section 7.3).

(5) **Conditional lethal mutations** These mutations are lethal under one set of conditions (the **restrictive conditions**) but permit normal growth under a different set of conditions (the **permissive conditions**). These mutations are very important, because they permit the analysis of essential genes in haploid organisms (particularly in bacteria and phages), as the mutant organism can be propagated under the permissive conditions and the effect of the mutation can be studied by using the restrictive conditions (Question 11.2).

(6) **Resistance mutations** Mutations which confer resistance to a particular drug or virus.

However, remember that these classes are somewhat notional and not mutually exclusive; for example, since most gene products catalyse specific biochemical reactions, nearly all gene mutations are, in reality, biochemical mutations.

11.3   The Effects of Mutation on DNA Structure

Gene or point mutations involve a single nucleotide pair (or occasionally a few adjacent nucleotide pairs) within the DNA of a gene or control sequence. There are two types of point mutation which affect gene expression in very different ways (Section 11.4). These are:

(1) **Base substitution mutations** A particular base pair is replaced by a different base pair. These substitutions are known as **transitions**, when a purine replaces a purine and a pyrimidine replaces a pyrimidine, and **transversions** when a purine replaces a pyrimidine, and vice versa:

$$
\text{Transitions} \qquad
\begin{array}{c} A \\ T \end{array} \longrightarrow \begin{array}{c} G \\ C \end{array}
\qquad\qquad
\begin{array}{c} T \\ A \end{array} \longleftarrow \begin{array}{c} C \\ G \end{array}
$$

$$
\text{Transversions} \qquad
\begin{array}{c} A \\ T \end{array} \longleftarrow \begin{array}{c} C \\ G \end{array}
\qquad\qquad
\begin{array}{c} T \\ A \end{array} \longrightarrow \begin{array}{c} G \\ C \end{array}
$$

(2) **Frameshift mutations** When one or two base pairs have either been added to or deleted from the gene.

11.4   The Effects of Point Mutation on Protein Synthesis

Consider mutations occurring within a short coding sequence from an imaginary gene (Figure 11.1A); four types of point mutation can be recognised, each of which exerts its effect in a different way.

(1) Same-sense mutations Although mutation has changed the nucleic acid sequence of a codon, the altered codon still encodes the same amino acid; this is possible because the code is degenerate (Section 9.3c). Both GAU and GAC are codons for aspartic acid, so that a GC for AT substitution in the third position of the DNA codon (Figure 11.1B) does not result in an amino acid substitution in the polypeptide. Thus same-sense mutations are silent.

(2) Missense mutations A single base pair substitution changes a codon for one amino acid into a codon for a different amino acid, resulting in an amino acid substitution in the polypeptide. In Figure 11.1(B) an AT to GC transition in the

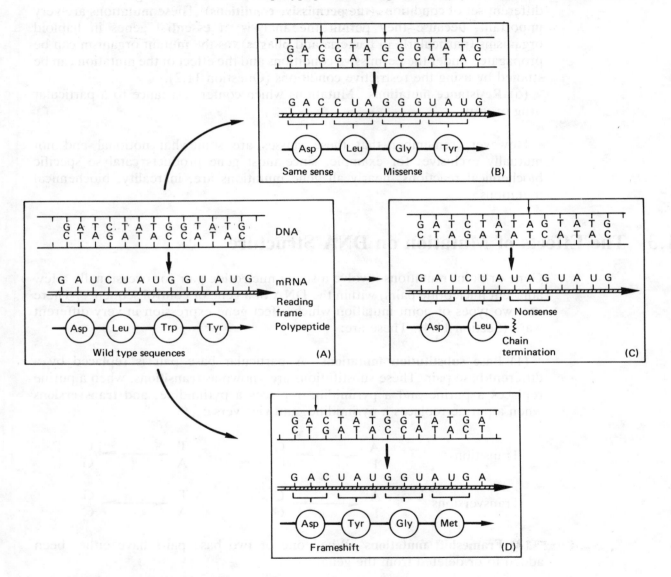

Figure 11.1 *Point mutations. The figure shows how a short nucleotide sequence within a gene is transcribed into mRNA and translated into a polypeptide.*
(A) The wild type sequences.
(B) A same-sense mutation has occurred in the aspartic acid codon and a missense mutation in the tryptophan codon.
(C) A nonsense mutation has occurred in the tryptophan codon.
(D) A (−1) frameshift mutation in the aspartic acid codon

DNA converts a UGG (tryptophan) into a GGG (glycine) codon. Most missense mutations reduce or abolish gene expression by preventing the polypeptide from folding-up into its active three-dimensional configuration; however, some missense mutations are silent, as the amino acid substitution does not always critically affect the tertiary structure of the gene product.

(3) **Nonsense mutations**   A base pair substitution changes the codon for an amino acid into a UAG, UAA or UGA chain termination triplet (Figure 11.1C). This causes the premature termination of translation, so that only an inactive polypeptide fragment is produced (Section 9.4).

(4) **Frameshift mutations**   These are due to the addition or deletion of one or two base pairs (Figure 11.1D). This shifts the reading frame so that all codons from the one containing the mutation to the end of the message are read out of phase and encode incorrect amino acids (Questions 9.2 and 9.3). Usually, however, the change in position of the reading frame generates an in-phase nonsense codon somewhere between the frameshift and the end of the gene, and this, in turn, results in the premature termination of translation.

Mutations may also occur in the genetic regulatory sequences. For example, a base substitution in the Pribnow box of a promoter may reduce promoter efficiency, while the addition or deletion of a base pair may alter the critical spacing between the Pribnow box and the transcriptional initiation site: these mutations in the regulatory sequences only affect the *rate* of gene expression and they only affect genes located in the same operon and on the same molecule of DNA. Mutations in regulatory sequences are described in Chapter 10.

11.5   Spontaneous Mutation

The origin of spontaneous mutations is poorly understood but some are almost certainly due to occasional uncorrected errors introduced during replication or DNA repair. The cell contains several mechanisms for repairing mismatched bases (either those escaping the proof-reading activities of the DNA polymerases or those introduced into the DNA after it has replicated) and other types of damage to its DNA; although these processes are normally both efficient and accurate, some errors do escape the correction process and lead to mutation.

Other spontaneous mutations are due to **tautomeric shifts**. In addition to the common form, each of the four bases can exist in a rare alternative or tautomeric form due to the rearrangement of some of the H atoms in the molecules; these rare **tautomers** have altered base pairing properties. For example, the rare enol form of thymine pairs with guanine, whereas the normal keto form pairs with adenine. Thus, if a shift to the enol tautomer occurs *in situ* then, at the next replication the enol tautomer ($\overset{*}{T}$) will pair with G (Figure 11.2). By the next replication the enol tautomer will probably have reverted to the normal keto form and normal base pairing will be resumed — the result is an AT → GC transition. Similarly, if the enol tautomer is incorporated into replicating DNA in place of C, the result will be a GC → AT transition.

The rates of spontaneous mutation for some representative procaryotic and eucaryotic genes are shown in Table 11.1. Note that in phages and bacteria an 'average' gene has a forward mutation rate of about 10^{-8} per bacterium per

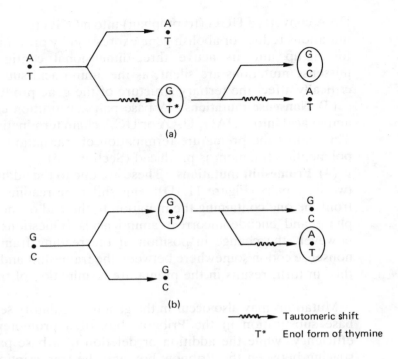

Figure 11.2 *Transitions caused by tautomeric shifts.*
(a) The keto form of thymine shifts in situ *to the enol form, resulting in an*
AT → GC transition.
(b) The enol tautomer is incorporated into replicating DNA in place of cytosine and
causes a CG → TA transition

generation, whereas in eucaryotes the mutation rates are rather higher — typically about 10^{-5} per gamete per generation.

11.6   Induced Mutation — the Molecular Basis of Mutagen Activity

The rate of mutation may be increased by exposing an organism to a **mutagen**, a chemical or physical agent which interacts with the DNA and, in one way or another, introduces errors at a subsequent replication. We shall consider how some commonly used mutagens exert their mutagenic effects.

(a)   Base analogues

A **base analogue** is a substance which can substitute for a normal base during DNA replication and, because they are correctly H bonded to a base on the template strand, they are not excised by the proof-reading activity of the DNA polymerases. They are weak mutagens, and, in order to produce a reasonable number of mutations, bacteria must be grown in their presence for many generations.

2-aminopurine (2-AP) is an analogue of adenine and normally it pairs with thymine; occasionally, however, it can pair with cytosine by the formation of a

Table 11.1

Organism and character			Mutation rate	
Man				
haemophilia			3×10^{-5}	per gamete per generation
Duchenne muscular dystrophy			$4\text{--}10 \times 10^{-5}$	
Mus musculus — mouse				
dilute coat	$d^+ \rightarrow$	d	3×10^{-5}	
pink eye	$p^+ \rightarrow$	p	9×10^{-4}	
Zea mays — maize				
non-purple aleurone	$Pr \rightarrow$	pr	1×10^{-5}	
shrunken endosperm	$Sh \rightarrow$	sh	1×10^{-6}	
Drosophila melanogaster — fruit fly				
white eye	$w^+ \rightarrow$	w	3×10^{-5}	
yellow body	$y^+ \rightarrow$	y	12×10^{-5}	
Neurospora crassa				
adenine independence	$ad3 \rightarrow$	ad^+	$* 4 \times 10^{-8}$	per conidium
Escherichia coli				
histidine requirement	$his^+ \rightarrow$	his^-	2×10^{-6}	per bacterium per generation
histidine independence	$his^- \rightarrow$	his^+	$* 1 \times 10^{-9}$	
lactose non-fermenting	$lac^+ \rightarrow$	lac^-	2×10^{-6}	
lactose fermentation	$lac^- \rightarrow$	lac^+	$* 2 \times 10^{-7}$	

Note that forward mutation rates (asterisked) are expected to be higher than the corresponding back mutation rates.

single H bond. When 2-AP is incorporated into DNA in place of either A or G, it causes the transitions $AT \rightarrow GC$ and $GC \rightarrow AT$ (Figure 11.3). Although 2-AP is is only rarely incorporated into DNA, once incorporated it acts as a potent mutagen.

Another commonly base analogue is **5-bromouracil (5-BU)**, an analogue of thymine. The common tautomer pairs with adenine but the rarer enol form pairs with guanine. Thus, 5-BU, like 2-AP, induces both $AT \rightarrow GC$ and $GC \rightarrow AT$ transitions.

Because base analogues induce transitions in both directions, mutations induced by a base analogue can also be reverted by a base analogue.

(b) Intercalating Agents

These are flat three-ringed aromatic molecules with the approximate dimensions of a base pair and they can **intercalate**, or slide in between, two adjacent base pairs along the double helix. In some unknown way this causes errors at the next

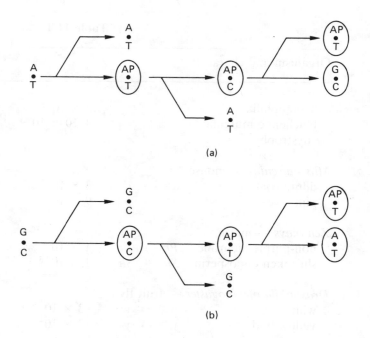

Figure 11.3 *Transitions induced by 2-aminopurine.*
(a) 2-AP is incorporated into DNA in place of adenine, leading to an AT → GC transition.
(b) The incorporation of 2-AP in place of guanine leads to a GC → AT transition

replication; most frequently one (or sometimes two) base pairs are added to the DNA sequence or, less frequently, a base pair is deleted. The result is a frameshift mutation.

Examples of intercalating agents are 9-aminoacridine and ethidium bromide. Note that intercalating agents do not induce either transitions or transversions.

(c) Nitrous Acid (HNO₂)

This is a very potent mutagen which acts by oxidative deamination. It causes mutations by converting adenine to hypoxanthine and cytosine to uracil; since hypoxanthine pairs with cytosine and uracil pairs with adenine, it induces both AT → GC and GC → AT transitions.

Occasionally deamination occurs spontaneously, so that some spontaneous mutations are probably due to this process.

(d) Alkylating Agents

These highly reactive compounds induce mutations by transferring an alkyl group (CH₃⁻, CH₃CH₂⁻, etc.) to one of the bases and changing its pairing specificity; thus, O⁶-alkylguanine can pair with thymine and O⁴-alkylthymine can pair with guanine, leading to AT → GC and GC → AT transitions. Furthermore, alkylating agents cause extensive damage to DNA by inducing strand breakage and depurination, too much for the normal repair systems to cope with. In this situation, when extensively damaged DNA is present, a further repair system is induced; however, this **inducible** or **SOS repair system** is very inaccurate (it is said to be **error-prone**), and many mismatched bases are introduced into the DNA during the repair process. The result is the production of many mutations, both transitions and transversions.

These are dangerous chemicals; not only are they highly toxic and, even at low concentrations, highly mutagenic, but many of them are also strongly carcinogenic. They include many chemicals of industrial importance and they are dangerous pollutants of our environment.

Alkylating agents used as mutagens include mustard gas, ethylmethane sulphonate (EMS) and *N*-methyl-*N*-nitro-*N*-nitrosoguanidine (NG), but, because they are so dangerous, they are no longer in general use.

(e) Electromagnetic Irradiation

Ultraviolet, X- and gamma-radiations are the radiations most frequently used in the experimental production of mutations. These radiations also occur naturally and are probably responsible for many of the spontaneous mutations.

Ultraviolet radiations, with a wavelength of about 254 nm, are strongly adsorbed by DNA and they exert their effect by the excitation of orbital electrons, raising them to higher energy states. Its principal result is the production of **pyrimidine** (usually thymine) **dimers**; two adjacent pyrimidines on the same DNA strand become crosslinked together and so prevent normal replication (Figure 11.4). At low doses the few dimers induced can be repaired by the normal repair systems of the cell, but as the dose increases (and, hence, the number of dimers), the normal repair processes are unable to cope and the SOS repair system is induced. Once again many mismatched bases are introduced into the DNA and result in the production of many mutations.

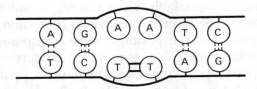

Figure 11.4 *A pyrimidine dimer. Covalent bonds form between two adjacent pyrimidines on the same strand; this distorts the double helix and weakens the H bonds between the strands*

Ultraviolet is a low-energy radiation and it is only weakly penetrating; although it is an effective mutagen in microbial systems, it is of very limited use in inducing germinal mutations in higher organisms.

X-rays and **gamma-radiations** are high-energy, deeply penetrating ionising radiations and exert their effect in two ways. First, quanta of energy hit the DNA and cause breaks in the sugar–phosphate backbone chains (the direct effect). Second, and more important, the radiations cause the ionisation of the constituent atoms of the molecules they meet; high-energy electrons are released, resulting in the production of highly reactive free radicals and ions which, in turn, cause numerous chemical changes in the DNA (the indirect effect). This damage induces the error-prone SOS repair system, and the end result is the production of many mutations.

The ionising radiations are highly mutagenic and the frequency of lesions is proportional to the dose administered. Because they cause extreme radiation damage, they should only be used with extreme care and only when no suitable alternative is available.

11.7 Transposable Genetic Elements

Transposable genetic elements are characterised by their ability to move or **transpose**, albeit at a low frequency, from one location in the genome to another. Since transposition occurs more or less at random, a transposable element will frequently insert into a structural gene or genetic regulatory sequence and cause a heritable change in gene expression.

Transposable elements were first postulated by Barbara McClintock in the late 1940s to account for unusual patterns of inheritance of pigment distribution on the grains and cobs of maize; during the following 20 years she demonstrated by painstaking genetic analysis that these elements, which she called **controlling elements**, could not only transpose but also suppress gene activity, cause localised mutagenesis, induce chromosome breakage at the site at which they were inserted and regulate gene activity during development. However, most contemporary geneticists rejected the concept of mobile genetic elements, and it was not until 1968, when molecular studies confirmed the existence of transposable elements in *Escherichia coli*, that the importance of such elements was realised.

Transposable genetic elements are naturally occurring parts of the genome of many organisms, both procaryotes and eucaryotes, and even in a simple organism such as *E. coli* there are many different transposable elements. In *E. coli* transposable elements are classified as either insertion sequences or transposons; **insertion sequences** (Question 11.7) only encode functions related to their own transposition, while **transposons** are much longer sequences which include at least one gene which confers upon the host bacterium a heritable property, most commonly resistant to a particular antibiotic (Question 11.8).

Each type of transposable element has a defined DNA sequence and can *only* exist when it is a part of an independent replicon, such as a bacterial chromosome or plasmid.

It is now realised that these elements are an important source of spontaneous mutations.

11.8 Mutagenicity Testing

Many cancers in humans are caused by exposure to toxic chemicals known as **carcinogens**; these cause normal cells to become malignant by dividing uncontrollably and spreading to other parts of the body, and they will induce tumours when injected into experimental animals. Once a number of carcinogens had been identified, it became clear that they induced DNA damage and that most were also mutagens; since carcinogenicity testing is such a slow and very expensive process, there was a need to develop alternative methods for detecting carcinogens. The best-known of these is the **Ames test**, developed by Bruce Ames in the 1970s, which detects potential carcinogens by assaying their effect on the reversion of histidine-requiring mutants of the bacterium *Salmonella typhimurium* (Question 11.6). Ames found that 95% of all the known carcinogens tested were also mutagenic and, correspondingly, that most of the substances identified as being mutagenic were also carcinogenic when tested on experimental animals.

These tests are important, as they enable substances that are mutagenic and, hence, potentially carcinogenic, to be rapidly and cheaply identified. Mutagenicity test systems have also been developed in rats and mice but these, like carcinogenicity tests, are very expensive and time-consuming.

Among the substances identified as being both mutagenic and carcinogenic are ionising radiations, many chemicals used in industry (for example, nitrosamines, aromatic amines and vinyl chloride), flame retardants, fungicides, insecticides, antibiotics and some chemicals formerly used in hair dyes or as food additives. A few exceptional substances, such as asbestos, are strongly carcinogenic but are not mutagenic.

11.9 Questions and Answers

Question 11.1

Explain what is meant by screening and show how it differs from selection.

Answer 11.1

Screening and selection are techniques of particular use in microbial systems. **Selection** uses a special set of environmental conditions which *only* permits the survival of mutant or recombinant cells with a particular phenotype. Examples of selection are the isolation of (i) drug- or phage-resistant bacteria by plating wild type cells on a medium containing a specific drug or phage and (ii) prototrophic reversions of an auxotroph (his^+ reversions of a histidine-requiring mutant, for example) by plating very many cells on minimal medium. For the most part, selective techniques only enable the detection of back mutations and they are useful in both isolating mutants and estimating the rates of mutation and recombination.

Screening methods are used to isolate mutants, frequently after treatment with a mutagen, and they facilitate the detection of rare mutants among a large number of wild type cells. Examples of screening are:

(1) The isolation of non-fermenting mutants on an indicator medium. For example, rare lac^- (lactose non-fermenting) mutants can be distinguished by plating several thousand wild type cells on a nutrient medium containing tetrazolium; at the high pH produced within a lac^- colony, the tetrazolium is reduced to dark-red formazan, whereas in lac^+ colonies the low pH inhibits this reaction and they remain whitish. Thus, the required lac^- colones are a dark

blood-red and are easily distinguished from the thousands of whitish *lac+* colonies.

(2) The enrichment technique used when isolating auxotrophs. In one method the wild type strain is grown for many generations in liquid minimal medium containing penicillin. Penicillin inhibits the growth of cell walls in dividing cells and so selectively inhibits the growth of wild type cells and, at the same time, increases the proportion of non-dividing auxotrophs. The auxotrophs can be detected by plating several thousand bacteria on minimal medium containing a small quantity of the required growth factor(s). If, for example, histidine is added, then the wild type *his+* cells will grow into large colonies, while any *his−* mutants will only form pinhead-sized colonies, because of the limiting amount of histidine in the medium.

Thus, screening techniques are used to increase the probability of isolating rare mutants.

Question 11.2

Mutations in a structural gene can be reverted by a back-mutation which restores the original wild type codon. Describe two other types of mutation which can partially or wholly restore the wild phenotype.

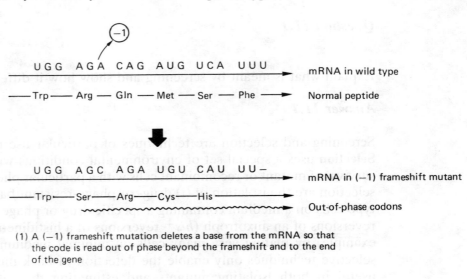

(1) A (−1) frameshift mutation deletes a base from the mRNA so that the code is read out of phase beyond the frameshift and to the end of the gene

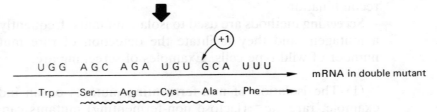

(2) If a (+1) frameshift mutation now adds a base to the mRNA, then the only codons misread are those lying between the two frameshift mutations

Figure 11.5 *Intragenic suppression*

Answer 11.2

The two mechanisms involve a second mutation, which may be located within the same gene as the original forward mutation (intragenic suppression) or within a different gene (intergenic suppression).

Intragenic suppression An excellent example is Crick's work on the *rIIB* mutants of phage T4 (see Question 9.2). Some *rIIB* mutants are due to a (-1) frameshift mutation within the *rIIB* coding sequence, so that the code is read out of phase beyond the frameshift (Figure 11.5.1).

If a further mutation (a ($+1$) frameshift) occurs at a nearby site, then only the codons between the two frameshifts will be read out of phase; these misread codons will cause amino acid substitutions in the polypeptide, but, provided that these substitutions are acceptable, a wild or near-wild type polypeptide will be produced (Figure 11.5.2). Thus, the second mutation suppresses the expression of the original frameshift. Note that in this type of suppression each mutation on its own will be mutant, but when they are combined into the same gene, a wild phenotype is produced.

Intergenic suppression This enables a missense or nonsense mutation (and certain frameshifts) to have a near wild phenotype if a particular suppressor mutation in a different gene is present. These suppressor mutations occur in genes encoding one or other of the transfer RNA molecules.

Suppose that mutation generates a UAG nonsense codon within the *rIIB* gene (for convenience, the nucleotide sequences are represented by the nucleotides present along the single strand of messenger RNA); translation will terminate prematurely at this codon and an incomplete and inactive polypeptide will be produced (Figure 11.6.1). These phage genes are translated by the bacterial translational machinery and there is no species of normal tRNA that recognises a nonsense codon. However, if another mutation (this time in a bacterial gene) occurs within the anticodon of a tRNA gene, it may produce a new anticodon (in this example, CUA) which can recognise and base pair with the UAG nonsense codon. The mutated tRNA is still charged with its own amino acid and so this amino acid will now be inserted in response to the nonsense codon; a complete polypeptide is produced and, so long as the inserted amino acid is acceptable, the polypeptide will have at least partial activity (Figure 11.6.3).

This is known as **nonsense suppression**. **Missense suppression** is similar, but the mutant tRNA now inserts an acceptable amino acid in place of an unacceptable one.

NOTES

1 Missense mutations can also be suppressed by intragenic suppression.
2 The nonsense mutants are an important type of conditional lethal mutant, as the mutation is not expressed in strains carrying the appropriate suppressor mutation (the permissive strain).

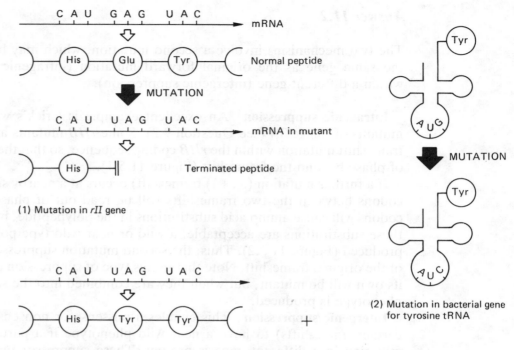

Figure 11.6 *Intergenic (nonsense) suppression.*
(1) Mutation generates a UAG chain termination codon within the rII *gene of phage T4.*
(2) Another mutation in the bacterial gene encoding one of the species of tyrosine tRNA alters the GUA anticodon to CUA.
(3) When both mutations are present in the same cell, the mutated tRNA can insert tyrosine in response to the UAG codon

Question 11.3

What are the constraints which determine whether or not a particular nonsense mutation will be suppressed by a given suppressor mutation? Why is it unlikely that a suppressor will be found which inserts phenylalanine in response to a nonsense codon?

Answer 11.3

The constraints are summarised as follows:

(1)   The suppressor tRNA must be able to read the particular nonsense codon as sense. This is usually the result of a mutation altering the anticodon of a species of tRNA, but rarely it may involve a nucleotide substitution in another part of the tRNA molecule.

(2) The anticodons on the normal and suppressor tRNA (or the codons read by the normal tRNA and the nonsense codon) must differ *only* by a single base

substitution — otherwise more than one mutation is necessary to generate the anticodon on the suppressor tRNA.

Thus, amber (UAG) mutations can only be suppressed as a result of a mutation in the gene encoding the tRNA molecule for one or other of the amino acids glutamic acid (normal anticodon CUC), lysine (CUU), glutamine (CUG), serine (CGA), tryptophan (CCA), tyrosine (AUA) and leucine (CAA).

(3) The suppressor tRNA must insert an acceptable amino acid. A substitution that is 'sense' in one polypeptide may be 'missense' in a different polypeptide. Thus, a particular nonsense suppressor will only suppress certain nonsense mutations.

(4) The suppressor mutation must not be lethal to the cell. Not only may the suppressor tRNA suppress nonsense mutations, but also it may read as sense some naturally occurring chain termination triplets. This is one reason why most suppressor mutations result in decreased viability. Furthermore, if, for example, an amber suppressor has arisen as a result of mutation in a gene for tyrosine tRNA, the cell *must* continue to insert tyrosine at the naturally occurring UAU and UAC codons. This is possible, because many tRNAs are encoded by more than one gene, so that a suppressor mutation in one tRNA gene does not affect the properties of the tRNA or tRNAs encoded by the other gene copies.

The codons for phenylalanine are UUU and UUC, read by the anticodons AAA and GAA. Amber is UAG and can only be read by a CUA anticodon. Thus, two separate mutations are required to convert either AAA or GAA to CUA.

Question 11.4

Write a short account of mutator genes in *Escherichia coli*.

Answer 11.4

Mutator genes are genes within which certain mutations occur and result in an increased frequency of mutation in the other genes; these mutations *must*, therefore, affect proteins that are required for accurate replication and they can increase the frequency of mutations by several orders of magnitude. For example, *mutT1*, discovered by Howard Treffers in 1954, increases the frequency of reversion of *trpA23* to wild type from 5×10^{-9} to 2.1×10^{-5}, a $\times 4200$ increase. Later work by Charles Yanofsky has shown that *mutT1*-induced mutations are specifically A–T to G–C transversions. It is not known how *mutT1* exerts its effect.

Subsequently, other mutators have been identified (not only in procaryotes, but also in yeast and higher eucaryotes) and shown to act by reducing either the accuracy of replication or the efficiency with which mismatched bases are corrected. A good example is the *mutD* mutations which occur within *polC*, the gene encoding one of the subunits of DNA polymerase III. These mutations reduce the $3' \rightarrow 5'$ exonuclease activity of DPIII, so that the mutant strains cannot effectively carry out the proof-reading process; hence, the mismatched bases introduced during replication cannot be excised and replaced by the activity of DPIII. Phage T4 encodes its own DNA polymerase and certain gene 43 mutants also have reduced exonuclease activity and increased frequencies of mutation; conversely, other gene 43 mutants have increased exonuclease activities, and these have lower than normal rates of mutation (these are **antimutator** mutants).

Other mutator genes have impaired abilities to carry out mismatch repair, a process that scans newly replicated DNA for any mismatched base pairs that have escaped the proof-reading process.

NOTE

It is estimated that newly replicated DNA contains about 1 in 10 000 (10^{-4}) mismatched bases. About 1 in 1000 of these escape the proof-reading process (10^{-7} overall) and only about 1 in 1000 of these fail to be corrected by mismatch repair (10^{-10}). It is these uncorrected mismatches that give rise to base substitution mutations.

Question 11.5

Write a short essay on mutational hot spots.

Answer 11.5

Although mutation is described as a random process, this is not quite true, as some sites (that is, base pairs) within a gene are more susceptible to mutation than others. Since there are only two base pairs (A–T and C–G), the frequency of mutation at these highly mutable sites, or **hot spots**, must be determined by the adjacent nucleotide sequences.

Hot spots were first detected by Seymour Benzer in 1961. He isolated over 2400 spontaneous *rII* mutants of phage T4 and mapped the site of each mutation within the *rIIA* and *rIIB* genes; altogether he detected 288 different sites. At about one-half of the sites only a single mutation was observed and most of the other sites had mutated between 2 and 20 times (that is, many of the mutations were identical and occurred 2 or more times) but there were two exceptional sites; one had mutated 312 times and the other 615 times. Mutant hot spots were also detected after treatment with chemical mutagens, but these hot spots were at different sites compared with the hot spots for spontaneous mutation.

More recently J. H. Miller has studied the distribution of amber mutations within the *E. coli lacI* gene. Amber mutations can be generated at 37 sites within *lacI* as the result of single base pair substitutions, and two of these sites are hot spots, accounting for about 40% of all *lacI* amber mutations. Both these hot spots fall within the same symmetrical sequence

$$5' \quad C \quad \overset{*}{C} \quad A \quad G \quad G \quad 3'$$
$$3' \quad G \quad G \quad T \quad \underset{*}{C} \quad C \quad 5'$$

196

This is the target sequence for the enzyme cytosine methylase, which methylates both asterisked cytosines in this sequence wherever it occurs along a DNA molecule. This converts cytosine to 5-methylcytosine. In wild type cells cytosine is occasionally spontaneously deaminated to uracil, but these uracils are rapidly excised from the DNA by another enzyme. However, when 5-methylcytosine is deaminated, it is converted to 5-methyluracil, better known as thymine and, since this is a naturally occurring base in DNA, it cannot be excised from the DNA like uracil. The result is a G–T mismatched base pair which, at the next replication, will produce an A–T for G–C transition. Thus, every 5-methylcytosine is potentially mutagenic.

NOTE

The experiments of Benzer are one of the most remarkable and most detailed fine-structure mapping experiments ever carried out. At the time they provided the most convincing demonstration that the gene was not an indivisible unit but consisted of many different sites separable by both mutation and recombination.

Question 11.6

Describe how potentially carcinogenic chemicals can be detected by their mutagenic activity in bacteria.

Answer 11.6

Since most mutagens are also carcinogens, any compound found to be mutagenic has a high probability of being a carcinogen. The simplest and most sensitive test for detecting chemical mutagens in bacteria is the Ames test.

This test uses a set of four different histidine-requiring (*his*⁻) tester strains of *Salmonella typhimurium*. Each *his*⁻ tester reverts to *his*⁺ in response to a different type of base substitution or frameshift mutagen, and the sensitivity of each tester is increased by two further mutations, one of which increases the permeability of the cell to foreign molecules, while the other prevents the changes induced by the mutagen from being repaired by naturally occurring repair systems.

In a test, between 10^8 and 10^9 cells of each tester strain are spread on minimal medium plates, a disc of filter paper saturated with a solution of the mutagen is placed in the centre of each plate and the plates are incubated for 2 days. The *his*⁻ bacteria cannot grow, but each mutation to *his*⁺ results in the formation of a small colony of *his*⁺ bacteria. If the compound under test is mutagenic, the disc on at least one plate will be surrounded by a halo of *his*⁺ colonies; the more potent the mutagen the greater the number of colonies (Figure 11.7). As controls the four testers are spread on minimal medium plates without the addition of a potential mutagen; this enables the frequency of spontaneous mutation to be estimated.

In practice the minimal medium plates also contain (1) a small amount of histidine, which enables one or two rounds of replication and is important

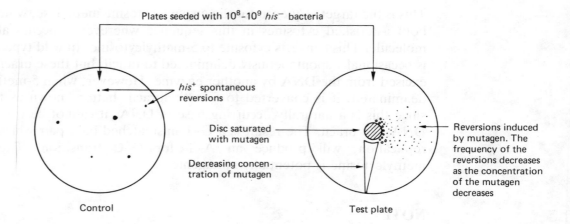

Plates seeded with 10^8-10^9 his^- bacteria

his^+ spontaneous reversions

Disc saturated with mutagen

Decreasing concentration of mutagen

Reversions induced by mutagen. The frequency of the reversions decreases as the concentration of the mutagen decreases

Control

Test plate

Figure 11.7 *The Ames test*

because some mutagens only act on replicating DNA, and (2) an extract from rat liver cells (the microsomal fraction). The latter is necessary because many potential carcinogens are activated by the hydroxylase systems present in liver cells but absent in bacteria; thus, the actual mutagen or carcinogen is often the metabolic product of an ingested chemical which may not itself be directly harmful.

This technique is of considerable practical importance, as it permits the rapid screening of suspected carcinogens that act by inducing point mutations. Not only is it a very sensitive test, but also it is quick (2–3 days) and very cheap.

Question 11.7

Write a short account of insertion sequences in *E. coli* and indicate their importance as a source of spontaneous mutation.

Answer 11.7

Insertion sequences (IS) are transposable genetic elements and they are only found inserted into self-replicating DNA molecules such as bacterial and phage chromosomes and plasmids.

In *E. coli* there are several different IS elements, typically between 0.7 and 1.5 kb long. Each has a defined nucleotide sequence and just one (or possibly two) genes encoding a transposase, a protein that promotes the transposition of that particular IS. In common with all transposable elements, the two ends of each IS are a pair of short inverted repeats, 9–41 bp long (Figure 11.8), and it is these IR sequences which the transposase apparently recognises in order to initiate transposition; thus, both the specific transposase and the specific IR sequences are absolutely essential for transposition. Another feature of transposition is that each element is flanked by a short (3–13 bp) direct repeat of host DNA; this sequence is the target site on the host DNA and is duplicated during the transposition process.

198

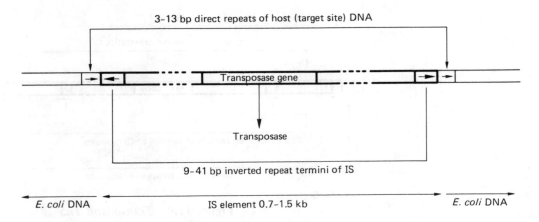

Figure 11.8 *The structure of* E. coli *IS elements*

When transposition occurs, the IS moves more or less at random to a new location in the genome. Frequently it is inserted into a structural gene and results in the appearance of a mutant phenotype, partly because the coding sequence has been interrupted and partly because IS elements contain a variety of both transcriptional and translational termination signals. Alternatively, the IS may be inserted into the operator–promoter region of an operon; this usually results in the entire operon being switched off but occasionally the operon is expressed constitutively. Constitutive expression results when the IS contains a correctly oriented promoter from which the bacterial operon can be transcribed — this promoter is not regulated by the normal regulatory protein for the bacterial operon and the effect is to mimic an operator-constitutive mutation.

Thus, the transposition of IS elements is an important source of spontaneous mutation. It is important to realise that these mutations can be neither induced nor reverted by base analogue and frameshift mutagens.

In *E. coli* there are several different IS elements with copy numbers between 1 and 15.

Question 11.8

Outline the structure and properties of bacterial transposons.

Answer 11.8

Transposons were discovered in 1974 following the observation that certain genes for antibiotic resistance appeared to be able to transfer from a plasmid to the bacterial chromosome, and vice versa.

Transposons are much larger than IS elements (typically 2–20 kb) and they have at least one gene which confers upon the host bacterium a heritable property, most commonly resistance to one or more antibiotics. This is a very useful property, as a plasmid can be 'tagged' with a transposon and the presence and transmission of the plasmid can be easily monitored by following the drug-resistant phenotype; likewise, transposition can be simply observed.

199

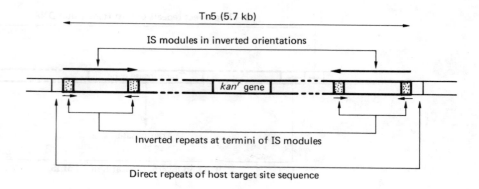

Figure 11.9 *Transposon Tn5*

Transposon Tn5 (Figure 11.9) is 5.7 kb long and is one of the simplest transposons; it is assembled from three components: a long central region (2.7 kb) containing a single gene for resistance to kanamycin, flanked by a pair of IS modules, each 1.5 kb long, in inverted orientations.

Other transposons are flanked by different pairs of IS modules and these are sometimes in the same orientation.

These transposons transpose like IS elements and duplicate a short sequence of host or target site DNA. Transposition can occur because, first, either one or both of the IS modules is functional and encodes a transposase (in some elements, including Tn5, one IS is only partly functional and does not encode an active transposase), and, second, the extremities of the transposon are always a pair of the inverted repeat sequences specific to the IS module: observe that these terminal inverted repeats are present whether the IS elements are in the same or in inverted orientations.

It is also probable that any pair of IS elements can co-operate to transpose any sequence lying between them, so that almost any gene can be made into a transposon by flanking it with two identical IS elements; this property has been exploited in the construction of recombinant DNA molecules.

NOTE

Tn5 is called a composite transposon because of its modular construction. Other transposons, the complex transposons, have a different organisation; they are flanked not by a pair of IS modules but just by a pair of inverted repeats, and the genes encoding the proteins required for transposition are located in the central part of the transposon.

Question 11.9

Describe the Muller-5 technique and show how it is used to assess the effect of X-irradiation on mutation frequency in *Drosophila melanogaster*.

Answer 11.9

In higher organisms a 'typical' gene has a forward mutation rate of 10^{-4}–10^{-5} per generation and it is extremely laborious to detect mutants and to measure accurately the rates of spontaneous and induced mutation of a *specific* gene. One method used to overcome this difficulty is the Muller-5 technique, devised by H. J. Muller. This does not measure the mutation rate of any one specific gene but measures the overall rate of mutation to all sex-linked recessive lethals at an unknown number of loci (perhaps 1000) on the X chromosome (Figure 11.10). Hence, the overall rate of mutation is sufficiently high to be measured accurately, enabling the effect of irradiation on mutation rate to be assessed.

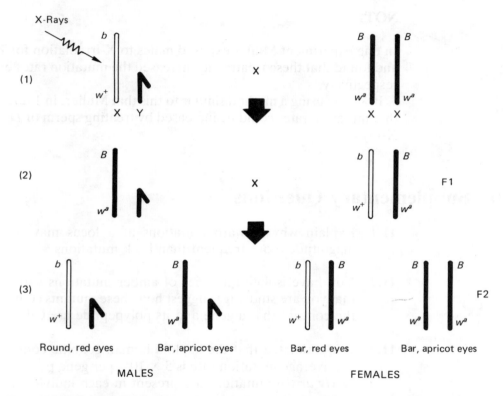

Figure 11.10 *The Muller-5 method for detecting sex-linked recessive lethal mutations in* Drosophila

The Muller-5 method is as follows.

(1) Wild type males are irradiated and mated with Muller-5 females; these females are homozygous for the Muller-5 X chromosome, which carries (i) the dominant *B* (bar-eye) gene which causes the eyes to be reduced to narrow slits; (ii) the recessive w^a (apricot eye) allele; and (iii) an inversion which prevents any crossing-over between the Muller-5 and the normal X chromosomes in heterozygous females.

(2) The F1 females are individually mated to F1 males. Note that every F1 female is heterozygous for a treated and an untreated Muller-5 chromosome, so that each F1 female represents one irradiated X chromosome.

201

(3) If no sex-linked recessive mutations are induced, there will be four types of progeny in a 1:1:1:1 ratio, but if the treated chromosome carries one or more sex-linked recessive lethal mutations, there will be **no** wild type males in the F2 culture and, hence, there will be twice as many females as males. Every such culture represents one mutation to an X-linked recessive lethal.

As a control the experiment is repeated using unirradiated males.

The effect of irradiation is assessed by comparing the frequency of recessive lethal mutations (i.e. the proportion of cultures with a 2:1 ratio of females:males) in the cultures derived from irradiated and unirradiated males.

NOTE

In one experiment Muller exposed males to X-irradiation for 24 min and 48 min and found that these treatments increased the mutation rates by ×9.5 and ×14.6, respectively.

It was by using a method similar to this that Muller, in 1927, first demonstrated that mutation rates could be increased by treating sperm of *D. melanogaster* with X-rays.

11.10   Supplementary Questions

11.1   Explain why forward mutations at a locus may be several orders of magnitude more frequent than back mutations.

11.2   You have isolated a series of amber mutations within a particular gene that you are studying. Suggest how these mutants could be used to confirm the concept that a gene and its polypeptide are colinear.

11.3   It is estimated that the haploid human genome contains 50 000 genes. If the average mutation rate is 5×10^{-5} per gene per generation, how many *newly arisen* mutations are present in each individual?

11.4   How can a missense mutation be suppressed by an intergenic suppressor mutation?

11.5   What is a tautomer and how may a tautomeric shift result in a mutation?

11.6   Why is it that mutations induced by the acridine dyes are more likely to be harmful to an organism than mutations induced by a base analogue?

11.7   Hydroxylamine is a highly specific mutagen when used on free phage or transforming DNA. It reacts only with cytosine in the DNA and converts it to a form that occasionally mispairs with adenine.

(a)   What base pair substitution is induced by hydroxylamine (HA)?
(b)   Would you expect mutations induced by HA to be reverted by it?

(c) Would you expect HA to revert mutations induced by 5-BU?

(d) Would you expect 5-BU to revert mutations induced by HA?

11.8 In which of the following genes could you isolate temperature-sensitive and/or amber mutations: (**a**) *lacO*; (**b**) *lacZ*; (**c**) *lacP*; (**d**) *lacI*?

11.9 Distinguish between back mutation, reversion and second site reversion.

11.10 Explain the difference between a mutation frequency and a mutation rate.

12 The molecular genetics of eucaryotes

12.1 Introduction

We have already described the overall organisation of the eucaryotic chromosome (Chapter 3) and in this chapter we shall consider some selected aspects of the molecular organisation of the DNA and its transcripts.

One curious feature is that eucaryotes appear to have far too much DNA! The human genome (3×10^9 bp) contains about 1000 times more DNA than does the *E. coli* chromosome, enough to encode between two and three million average-sized proteins, and yet 50 000 is a realistic estimate of the number of genes in a human haploid genome. Thus, only 5–6×10^7 bp, or 2% of the genome, is required for coding. Some animals and plants have even more DNA — up to 50 times the amount in humans. The function of much of this excess DNA is unknown.

12.2 Classes of Eucaryotic DNA

Eucaryotic genomes contain two fundamentally different classes of DNA, **unique DNA** and **repetitive DNA**.

The unique, or single-copy, DNA consists of unique, or nearly so, sequences and makes up about 70% of the human genome; however, less than 5% of this unique DNA can be accounted for by known coding and regulatory sequences. Some of the non-coding unique DNA is found as untranslated sequences within genes (Section 12.4) and some human genes may contain as much as 99% of non-coding DNA.

The repetitive DNA comprises a number of discrete sequences that are present many times. Sequences repeated between 20 and 10 000 times are referred to as **middle repetitive (MR) DNA**, while those repeated more than 5×10^4 times are termed **highly repetitive (HR) DNA**. Most of this repetitive DNA is non-coding.

Some examples of repetitive DNA are:

(1) **Redundant genes**   A few genes, including the tRNA, rRNA and histone-protein genes are **redundant**, as they are present in multiple copies (Section 12.6). These genes are part of the middle repetitive DNA, which makes up between 15% and 30% of the human genome.

(2) **The *Alu* sequences** These sequences, so-called because they have a sequence recognised by restriction endonuclease *AluI* (Section 13.3), are about 300 bp long and, in humans, there are about 3×10^5 copies dispersed around the genome (one *Alu* sequence per 10 kb of DNA). They account for about 3% of the human DNA. The *Alu* sequences are not identical but any two sequences have about 85% sequence identity.

The *Alu* sequence is only one of the HR sequences found in eucaryotic genomes; some sequences are much longer (5 kb or more).

(3) **Centromeric satellite (simple-sequence) DNA** These are short sequences, (usually less than 10 bp) repeated over 10^7 times per genome and found exclusively in the region of the centromeres. They probably play a role in aligning the chromosomes on the metaphase plate during cell division.

12.3   Transcription in Eucaryotes

In eucaryotes transcription is more complex than in procaryotes and differs in a number of important respects:

(1) Transcription and translation occur in different cellular compartments and are not coupled as in procaryotes. RNA transcribed in the nucleus must pass to the cytoplasm before it can function.

(2) All transcripts are synthesised as oversize precursor molecules.

(3) The precursor molecules are extensively processed. This includes the removal of any unwanted sequences from within the coding regions (Section 12.5).

(4) The precursor messenger RNA molecules have caps and tails added (Section 12.3d).

(5) In procaryotes genes of related function are usually clustered and transcribed onto a single polycistronic messenger. In eucaryotes every gene has its own promoter and is separately transcribed (except the rRNA genes).

(6) Many of the primary transcripts are very rapidly degraded by nucleases and never reach the cytoplasm.

(a)   The Primary Transcripts

The primary transcripts of the genes are mostly precursor molecules of messenger RNA and they collectively form the **heterogeneous nuclear RNA (hnRNA)**, a class of RNA found *only* in the nucleus. These precursor or pre-mRNA molecules are variable in size (up to 30 kb) and are much larger than the cytoplasmic messengers 0.5–3 kb). However, they contain all the sequences found in mature mRNA, are unstable and generally have both caps and tails (Section 12.3d); thus, they have all the properties of precursor mRNA molecules. The transcripts of the tRNA genes (**pre-tRNA**) and rRNA genes (**pre-rRNA**) are also synthesised as oversized molecules.

Before these precursor molecules can pass to the cytoplasm, they must be trimmed to size or **processed**. The processing of the pre-tRNA and pre-rRNA is similar to the processing of these molecules in procaryotes, as it involves the addition of some nucleotides and the removal of others and the chemical modification of bases (Section 8.3). In eucaryotes, however, the precursor RNAs, particularly the pre-mRNA, undergo a unique and essential type of processing — this is the removal of introns by **splicing** (Sections 12.4 and 12.5).

In eucaryotes the precursor RNAs are transcribed by **three** different RNA polymerases: RPI acts only in the nucleolus and transcribes the pre-rRNA; RPII transcribes the pre-mRNA; and RPIII transcribes the pre-tRNA and the 5 S rRNA. Each RNA polymerase recognises and binds to different promoter sequences (Question 12.3).

(b) The Promoters for RPII

In eucaryotes, as in procaryotes, there is a promoter which binds RPII and at which transcription is initiated. The promoters recognised by RPII all have a **TATA** or **Hogness box** (part of the consensus sequence TATA^A_TA^A_T) surrounded by GC-rich sequences and located about 30 bp upstream from the transcriptional initiation site. This is analogous to the Pribnow box in procaryotic promoters and, like the latter, probably functions by correctly aligning RPII relative to the intiation site.

Although the TATA box is necessary, it is not normally sufficient to initiate transcription and one or more upstream sequences are required. Many promoters also have a **CAAT box** (GGNCAATCT) and/or a **GC box** (GGGGCGG) located between −40 and −110 (Figure 12.1). These sequences, unlike the −35 sequence in procaryotes, can be on *either* strand of the DNA and their spacing from the TATA box is variable. This is because the upstream sequences are not themselves recognised by RPII but by specific proteins called **transcriptional factors**. Each type of upstream sequence is recognised by a different transcriptional factor and these complexes then promote the binding of RPII. Thus, the transcriptional factors and upstream sequences act to modulate the rate of transcriptional initiation.

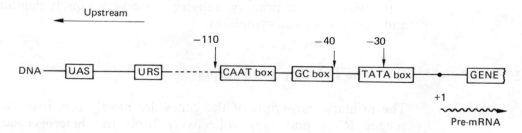

Figure 12.1 *Sequences found in promoters for RPII. The TATA box is nearly always present. The upstream sequences are variable, although at least one must be present for promoter activity. UAS and URS are not strictly within the promoter sequence*

Several hundred base pairs upstream there may be a further control sequence, either an **upstream activating sequence (UAS)**, a positive control sequence for switching on gene activity, or an **upstream repressing sequence (URS)**, a negative control sequence which acts to turn off transcription. These sequences are not always present.

(c) Termination

Very little is known about termination in eucaryotes. A few genes have a sequence resembling a procaryotic terminator, but more often termination is, at least in part, dependent on the transcribed sequence AAUAAA. This sequence is always present, although it may be hundreds of nucleotides from the point at which transcription is terminated. An endonuclease recognises this sequence and cuts the transcript some 20 nucleotides downstream from it (see Figure 12.2). If this sequence is missing, the transcript is extended, as termination does not occur.

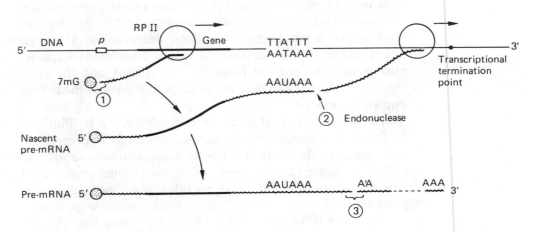

Figure 12.2 *Pre-mRNA is capped and tailed.*
(1) Soon after the initiation of transcription, a 7mG cap is added at the 5' end of the transcript.
(2) The AAUAA sequence is recognised by an endonuclease and the primary transcript is cut 20–30 nucleotides downstream.
(3) A poly(A) tail, consisting of 100–200 consecutive adenosine residues is added onto the newly exposed 3' end

(d) Caps and Tails

A further difference from procaryotic mRNAs is that all eucaryotic messengers have a 7-methylguanosine cap added at the 5' end and most (the exceptions being the mRNAs for the histone proteins and for some yeast genes) have a poly(A) tail added at the 3' end.

The cap is added very soon after the initiation of transcription (Figure 12.2). The 7-methylguanosine (7-mG) and the true terminal nucleotide are joined through their 5' sites; effectively, these nucleotides are joined back-to-back so that the cap blocks the 5' end of the message. Second, after the transcript has been cut beside the AAUAAA sequence, the new 3' end is polyadenylated by the addition of between 100 and 200 consecutive adenosine residues. The functions of the cap and tail are not clear but they do protect the message from degradation; furthermore, the caps must be present for ribosomes to bind to the messengers and initiate translation.

Transcription by RNA polymerase II is absolutely necessary for capping and tailing, so that neither tRNA nor rRNA is capped and tailed.

Once the cap and tail have been added, certain untranslated sequences within the coding sequences must be removed from the pre-mRNA by splicing (Section 12.5).

12.4 Split Genes

In bacteria and phages the genes are nearly always continuous and every nucleotide within a gene is a part of its coding sequence and is translated. However, in eucaryotes and their viruses most genes are discontinuous or **split**, and contain internal nucleotide sequences that do not appear in the mature RNA (and so are never translated). These **intervening (IV) sequences** are generally called **introns**, while the expressed (i.e. translated) sequences are **exons**. These introns must be cut out from the pre-mRNA and the remaining exons joined together before it can become mRNA and pass to the cytoplasm; this process is known as **splicing**.

Split genes were first discovered in adenovirus by Phillip Sharp in 1977, when he found that certain nucleotide sequences were present within a gene but did not appear in the mature mRNA. Almost simultaneously it was found that the mouse β-globin (a component of haemoglobin) gene and the chick ovalbumin (the major component of egg-white) gene were also split. The latter were particularly favourable systems to study, since large amounts of β-globin and ovalbumin mRNA can readily be purified from the special cells which produce them in large quantities.

These split genes were detected by (1) cloning and isolating pure genic DNA, (2) denaturing this DNA, (3) hybridising it with the corresponding pre-mRNA or mRNA, and (4) examining the DNA–RNA heteroduplex molecules by electron microscopy. In some experiments the DNA was completely denatured into its component strands and the hybrid molecules appeared as in Figure 12.3 (a); the hybrid molecules made from DNA and the pre-mRNA were paired uniformly all along their length, just as expected if the single strands carried precisely complementary nucleotide sequences, but when mRNA was used, one or more single-stranded loops was pushed out from the duplex structure — each loop, called an **R loop**, represents a sequence present on the DNA but absent from the mRNA. In other experiments the DNA was only partly denatured and the strands only separated just enough to allow hybridisation to be initiated; the experimental conditions were such that RNA–DNA hybrids formed in preference to DNA–DNA homoduplex molecules and the remaining section of each partially hybridised RNA strand then displaced the complementary DNA strand until heteroduplex formation was complete. The 'three-stranded' complex molecules (Figure 12.3b) again showed R loops whenever introns were present.

Observe that with a large intron the DNA in the R loop is able to reform a duplex structure along part of its length by reannealing with the complementary DNA strand.

The structure of the human β-globin gene is shown in Figure 12.4. This gene has two introns which, together, contain more than twice as many nucleotides as the coding sequence itself. Other genes may have as many as 60 introns and perhaps even more.

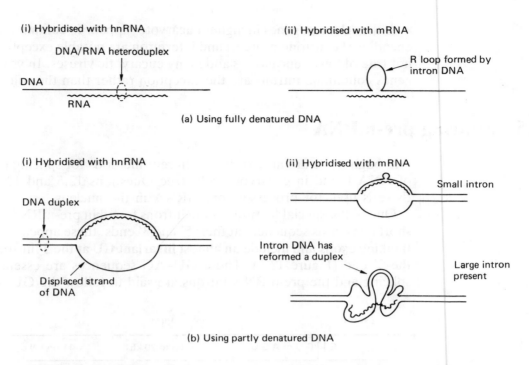

(a) Using fully denatured DNA

(b) Using partly denatured DNA

Figure 12.3 *The demonstration of a split gene. Genic DNA has been cloned, purified, denatured and hybridised with the corresponding species of hnRNA and mRNA. The hybrid molecules were treated with protein (this binds to them and effectively increases their diameter) and examined under the electron microscope. The interpretations of the observed structures show that sequences are missing from within the genic sequences of mRNA. R loops form where a sequence is present on the DNA and absent from the RNA*

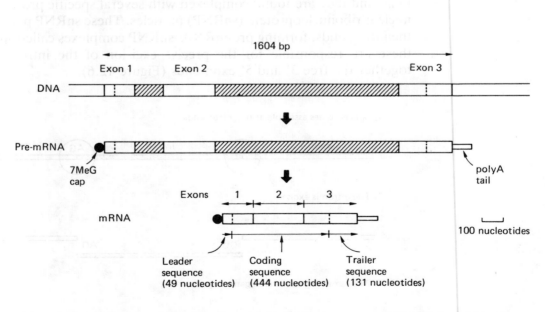

Figure 12.4 *The structure of the human β-globin gene, pre-mRNA and mRNA. The gene contains a small and a large intron (130 and 850 nucleotides, respectively) which are removed from the pre-mRNA by splicing*

209

Although most genes in higher eucaryotes appear to contain introns (the genes encoding the histone proteins and interferon are notable exceptions), the same is not true of lower eucaryotes and many eucaryotic viruses. In yeast, for example, genes containing introns are the exception rather than the rule.

12.5   Splicing pre-mRNA

Although there are at least four different methods for splicing the several types of RNA found in eucaryotic cells (see Questions 12.5 and 12.6), only one of these is used for processing pre-mRNA in the nucleus.

One of the special features of the introns found in pre-mRNA is that they have short consensus sequences at their 5′ and 3′ ends; these appear to extend into the flanking exons and include an almost **invariant GU** at the 5′ intron end and **AG** at the 3′ end (Figure 12.5). These GU–AG sequences are essential for accurate splicing and the pre-mRNA introns are said to follow the GU–AG rule.

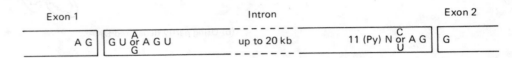

Figure 12.5   *Intron ends. The 5′ and 3′ ends of the introns found in pre-mRNA have almost invariant GU and AG sequences, respectively. These are part of longer consensus sequences*

The other requirement for splicing is the presence of several species of **small nuclear RNAs (snRNA)**; these are stable molecules only 100–215 nucleotides long, and they are found complexed with several specific proteins to form small nuclear ribonucleoprotein (snRNP) particles. These snRNP particles assemble at the intron ends, forming pre-mRNA–snRNP complexes called **spliceosomes**, and these are responsible for the precise excision of the intron and for joining together the free 3′ and 5′ exon ends (Figure 12.6).

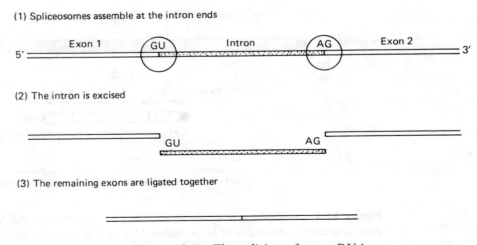

Figure 12.6   *The splicing of pre-mRNA*

Splicing must be an extremely accurate process, as if one too many or one too few nucleotides are excised, the resulting mRNA would contain a frameshift mutation and could not act as a template for a normal polypeptide. Furthermore, the splicing mechanism must ensure that the 5′ and 3′ ends of the same intron participate in the same splicing reaction. If, for example, the 5′ end of one intron and the 3′ end of a different intron participated in the same splicing event, then a large tract of coding RNA would be missing from the mRNA (Figure 12.7).

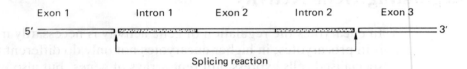

Figure 12.7 *Inaccurate splicing. A splicing event involving the 5′ end of intron 1 and the 3′ end of intron 2 will delete the whole of exon 2 from the mRNA*

12.6 Tandemly Repeated Genes

Most genes occur as single copies and there is only one copy of each gene per haploid genome, but some genes are remarkable, as they are tandemly repeated very many times. The best-known example is the genes encoding the three major types of RNA found in the eucaryotic ribosomes. The genes for the 18 S, 5.8 S and 28 S rRNAs occur as a cluster containing one copy of each gene, and this cluster is repeated in tandem 100–200 times per haploid genome. Each cluster is separated from the next by an untranscribed spacer (Figure 12.8). This tandem repetition enables the cell to synthesise the very large amounts of rRNA required for its ten million or so ribosomes.

Another distinctive feature of these rRNA genes is that each cluster is transcribed onto a single oversize precursor molecule, which is then processed, by removing the excess nucleotides between the different rRNA sequences, to form the three separate molecules of 18 S, 5.8 S and 28 S rRNA.

These tandemly repeated gene clusters are located in the nucleolar organisers, specialised regions of the chromosomes associated with the nucleoli.

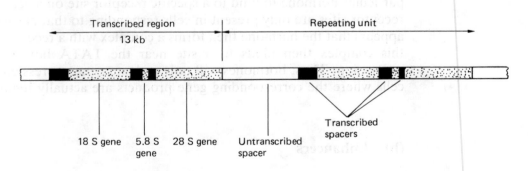

Figure 12.8 *Tandemly repeated gene clusters of mammalian rRNA genes. The cluster of three genes is repeated 100–200 times per haploid genome*

Other tandemly repeated genes are the 5 S rRNA genes, grouped in one or more clusters, depending on the species, and the gene cluster encoding the five histone proteins. Note that although the five genes encoding the histone proteins are clustered, each gene is separately transcribed.

12.7 Regulating Gene Activity

In eucaryotes the regulation of gene activity is necessarily more complex than it is in procaryotes; in higher eucaryotes not only do different tissues and different specialised cells express different arrays of genes, but also, throughout development from fertilised egg to adult, gene expression must be co-ordinated between different groups of cells, sometimes located in widely separated regions of the developing organism; there has to be a highly organised system of communication between these different cells.

As in procaryotes, regulation is primarily by the control of transcription, but because genes of related function are not clustered, each one must have its own promoter and regulatory sequences. The mechanisms of control are varied, often complex and poorly understood; we can only briefly indicate the diversity of these regulatory mechanisms.

A simple system, resembling operon control in procaryotes, regulates the expression of three genes required for the utilisation of galactose in yeast (Question 12.9). Each gene has its own promoter and upstream activating sequence (UAS) and each UAS is simultaneously stimulated to promote transcription by the binding of a positive regulatory protein encoded by an activator gene.

(a) Hormones and Gene Expression

In higher eucaryotes metabolism is largely regulated by **hormones**. How this hormonal regulation is achieved is poorly understood, but hormones do act by regulating the transcription of specific genes and, since they are a means of communication between cells, they are very important in development. The chain of events leading to the initiation of gene expression first requires a particular hormone to bind to a specific receptor site on a cell membrane; these receptor sites are only present in cells responding to that particular hormone. It appears that the hormone then forms a complex with a receptor protein and that this complex then binds to a site near the TATA box and activates gene expression. Thus, hormones will only activate gene expression in the particular cells where the corresponding gene products are actually required.

(b) Enhancers

Enhancers are regulatory sequences which stimulate transcription from eucaryotic promoters with which they are associated, and they are probably the major mechanism for regulating gene expression in eucaryotes (Question 12.8). They differ from most regulatory sequences, as they can be located several kilobases away from the promoter they control. Thus, their activity is difficult to explain in conventional terms and it is thought that they are stimulated by the binding of a specific protein, present only in those cells where gene activity is required, which enables the enhancer to act as a site for the assembly of the transcriptional initiation complex.

Among the proteins that are known to bind to activate enhancers are some hormonal receptor proteins, so it may be that enhancers or enhancer-like elements are the target sequences for the hormonal control of gene expression.

12.8   Questions and Answers

Question 12.1

You are given a 300 nucleotide long sequence believed to be that of a molecule of messenger RNA. How would you (a) confirm that this RNA was messenger and not either tRNA or rRNA and (b) decide whether it was a procaryotic or a eucaryotic messenger?

Answer 12.1

(a) It is too long to be tRNA. If it were a species of rRNA, it would contain (1) many modified bases such as pseudouridine and 5-methylcytosine and (2) inverted repeat sequences capable of forming hairpin loops. However, if it were mRNA, the sequence would include an open reading-frame consisting of an AUG initiation codon, a tract of in-phase amino acid codons and an in-phase chain termination triplet.

(b) All eucaryotic mRNAs have a 5′ 7-methylguanosine cap and most have a long 3′ polyA tail. These features are absent from procaryotic messengers, which, however, have a ribosome-binding (Shine–Delgarno) sequence near their 5′ ends.

Question 12.2

Tabulate the principal structural differences between the ribosomes of *E. coli* and a higher eucaryote.

Answer 12.2

	E. coli	Higher eucaryote
S value	70 S	80 S
Size	30 × 20 nm	32 × 22 nm
Larger sub-unit		
Size	50 S	60 S
Components	23 S rRNA (2904 bases)	28 S rRNA (4718 bases)
	5 S rRNA (120 bases)	5.8 S rRNA (160 bases)
		5 S rRNA (120 bases)
	c. 34 polypeptides	c. 49 polypeptides
Smaller sub-unit		
Size	30 S	40 S
Components	16 S rRNA (1541 bases)	18 S rRNA (1874 bases)
	c. 21 polypeptides	c. 33 polypeptides

Question 12.3

Why are the promoters recognised by RPIII so unusual?

Answer 12.3

RPIII transcribes a special set of genes with the following features:

(1) the transcripts are less than 300 nucleotides long;
(2) the transcripts never encode proteins;
(3) the genes are usually present in multiple copies.

This set of genes includes the 5 S rRNA and the tRNA genes.

The promoters for these genes, unlike the promoters for any other RNA polymerase, are located within the genes themselves. This was first recognised when it was found that all sequences located 5' to a 5 S rRNA gene, and even the first 50 nucleotide pairs of the gene itself, could be deleted without affecting the initiation of transcription; likewise, the deletion of sequences beyond +84 had no effect on transcriptional initiation. Thus, the sequence between +50 and +84 defines the promoter for the 5 S rRNA gene.

Even more unusual is that in the tRNA genes the promoter is split into two parts, the A and B blocks, and the intervening sequence can be deleted without affecting promoter efficiency.

Question 12.4

Describe how pre-mRNA is spliced to form messenger RNA.

Answer 12.4

Most eucaryotic genes contain one or more introns and the splicing mechanism most excise these introns and splice together the exons to form a messenger. The splicing process must be very precise, as the removal of too many or too few nucleotides would interfere with the correct reading of the code.

Consider a molecule of pre-mRNA containing a single intron (Figure 12.9.1). The 5' end of the intron starts with an invariant (almost) GU and the 3' end terminates with an AG; these dinucleotides are critical but they are part of longer consensus sequences.

Small RNPs (molecules of small nuclear RNA together with associated proteins) assemble at each of the consensus sequences and form the spliceosomes required for splicing. The RNA is now cleaved at the 5' GU site, between the last nucleotide of the exon and the GU (Figure 12.9.3), the intron RNA is

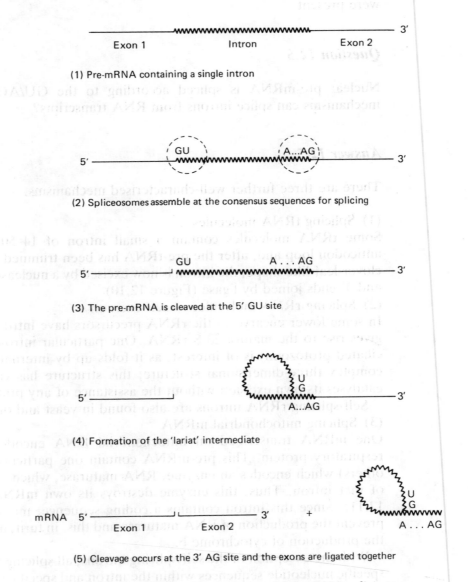

(1) Pre-mRNA containing a single intron

(2) Spliceosomes assemble at the consensus sequences for splicing

(3) The pre-mRNA is cleaved at the 5' GU site

(4) Formation of the 'lariat' intermediate

(5) Cleavage occurs at the 3' AG site and the exons are ligated together

Figure 12.9 *The lariat model for RNA splicing*

215

looped around and the 5′ terminal G is joined to an A within but close to the end of the intron (Figure 12.9.4). This involves the formation of a 2′–5′ phospho-diester bond, so that this A is connected to *three* other nucleotides and is a 'branched' nucleotide. Finally, cleavage occurs at the 3′ end of the intron (between the terminal U and the first nucleotide of exon 2) and exons 1 and 2 are ligated to form the messenger.

NOTE

This 'lariat' model for splicing was developed by Philip Sharp. He isolated a pure species of pre-mRNA containing a single intron and added this to cellular extracts. In this *in vitro* system splicing took place and the extracts could be isolated and examined. Further, he found that splicing only occurred if snRNPs were present.

Question 12.5

Nuclear pre-mRNA is spliced according to the GU/AG rule. What other mechanisms can splice introns from RNA transcripts?

Answer 12.5

 There are three further well-characterised mechanisms:

(1) Splicing tRNA molecules
Some tRNA molecules contain a small intron of 14–50 nucleotides in the anticodon loop and, after the pre-tRNA has been trimmed to size, it assumes a clover-leaf-like shape; the intron is now excised by a nuclease and the exposed 5′ and 3′ ends joined by ligase (Figure 12.10).
(2) Splicing rRNA molecules — self-excision
In some lower eucaryotes the rRNA precursors have introns in the region that gives rise to the mature 23 S rRNA. One particular intron in *Tetrahymena* (a ciliated protozoon) is of interest, as it folds up by internal base pairing into a complex three-dimensional structure; this structure has catalytic activity and catalyses its own excision without the assistance of any protein.
    Self-splicing rRNA introns are also found in yeast and other fungi.
(3) Splicing mitochondrial mRNA
One mRNA transcribed on mitochondrial DNA encodes cytochrome b, a respiratory protein. This pre-mRNA contain one particular intron (there are others) which encodes an enzyme, RNA maturase, which catalyses the splicing of that intron. Thus, this enzyme destroys its own mRNA template (Figure 12.11). Since this intron contains a coding sequence, mutations within it may prevent the production of RNA maturase and this, in turn, prevents splicing and the production of cytochrome b.
    With the exception of the self-splicing introns, all splicing mechanisms require specific nucleotide sequences within the intron and specific protein molecules to promote their excision.

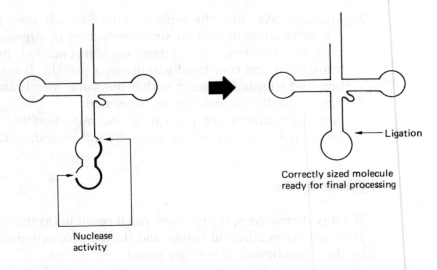

Ligation

Correctly sized molecule
ready for final processing

Nuclease
activity

Figure 12.10 *Intron splicing in yeast tRNA. The intron is represented by the heavy line*

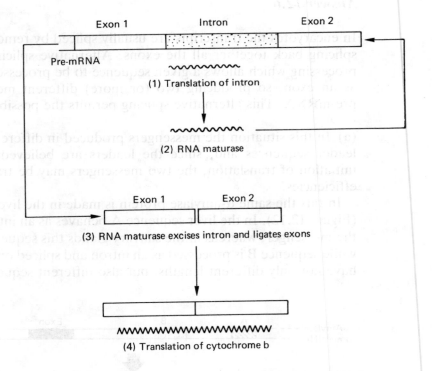

Exon 1   Intron   Exon 2

Pre-mRNA

(1) Translation of intron

(2) RNA maturase

(3) RNA maturase excises intron and ligates exons

Exon 1   Exon 2

(4) Translation of cytochrome b

Figure 12.11 *The splicing of yeast mitochondrial RNA*

NOTES

1 The self-splicing intron in *Tetrahymena* was described by Thomas Cech in
   1986, and provided the first demonstration that RNA could be an enzyme
   (called a **ribozyme**).

2 Mitochondria, like chloroplasts, contain their own genetic systems. Each organelle contains several small molecules of organellar DNA; this DNA, with the assistance of certain nuclear encoded proteins, is replicated, transcribed and translated within the organelle. It is only because transcription and translation occur within the same membrane-bound compartment that organellar introns can be translated.

3 Although introns are present in the mitochondrial DNA of many lower eucaryotes, they are absent from the mitochondrial DNA of mammals.

Question 12.6

What is alternative splicing? How can it result in (**a**) the differential regulation of gene activity in different tissues and (**b**) the production of two different proteins by the transcription of a single gene?

Answer 12.6

In eucaryotes the transcripts are usually spliced by removing all the introns and splicing back together all the exons. Alternative splicing is a variation in this processing which allows a given sequence to be processed either as an intron or as an exon, so producing two (or more) different messengers from a single pre-mRNA. This alternative splicing permits the possibility of gene regulation.

(**a**) In this situation the messengers produced in different tissues have different leader sequences and, since the leaders are believed to be involved in the initiation of translation, the two messengers may be transcribed with different efficiencies.

In rats the same α-amylase protein is made in the liver and the salivary gland (Figure 12.12). In the liver sequence A behaves as an intron and is spliced out of the messenger, whereas in the salivary glands this sequence behaves as an exon, while sequence B is processed as an intron and spliced out. Thus, the two leaders have not only different lengths, but also different sequences.

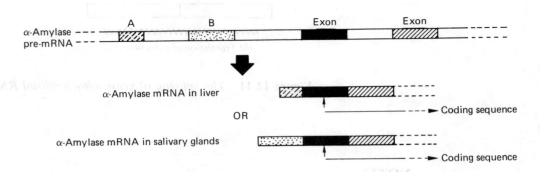

Figure 12.12 *Alternative splicing of the α-amylase RNA in rats. In the liver, sequence A (50 nucleotides long and known as exon L) is retained and sequence B is spliced out. In the salivary glands, sequence B (161 nucleotides long, exon S) is retained, while A is spliced out. The small arrow indicates the AUG translational initiator*

(b) In this situation the sequence or sequences involved are removed from within the coding sequence of a gene, so that alternative splicing (usually occurring in different tissues) can produce two varient forms of the same protein or even two proteins with very different functions.

This situation has been found in, for example, the calcitonin gene of mammals and birds. In the thyroid gland all the exons are present and calcitonin is synthesised, but in the neural tissue a six-nucleotide sequence is removed by alternative splicing and a shorter related peptide (calcitonin-gene-related peptide, CGRP) is found.

Question 12.7

How was it first shown that the consensus sequences at the ends of introns are essential for normal splicing?

Answer 12.7

Although it was comparatively simple to show that there were consensus sequences at the ends of introns, it was much more difficult to demonstrate their importance in normal splicing. One approach was to collect mutant genes with presumed splicing defects and to examine the altered sequences.

Humans suffering from the genetic disease β-thalassaemia have abnormally low levels of β-globin, one of the polypeptide sub-units of haemoglobin and encoded by the β-globin gene. In some patients there is no functional β-globin mRNA in the cytoplasm, although the pre-mRNA appears to be present in the nuclei; in these patients the defect is due to inaccurate splicing, a consequence of mutation.

These mutations were found to be of two types.

(1) A mutation which altered the consensus sequence at the 5′ end of intron 1:

5′ terminal sequence of intron 1 RNA   C A G G U U G G U
5′ sequence after mutation   C A G A U U G G U

This mutation destroys *normal* splicing; however, splicing is not abolished, as the remaining 3′ normal sequence now seeks out an alternative site with a sequence similar to the 5′ consensus sequence but at which splicing does not normally occur. There are several of these **cryptic** sites:

A A G G U G A A C ⎫
G U G G U G A G G ⎬   within exon 1
A A G G U U A G A   within intron 1

and often several different splices are made in the same mutant, so that a variety of abnormal polypeptides is produced (Figure 12.13a).

(2) Mutations within an exon or intron which generate a new splice junction in the RNA. Mutation in intron 2 created the new splice site C A G G T A C C A (Figure 12.13b), so that missplicing effectively creates a new exon.

219

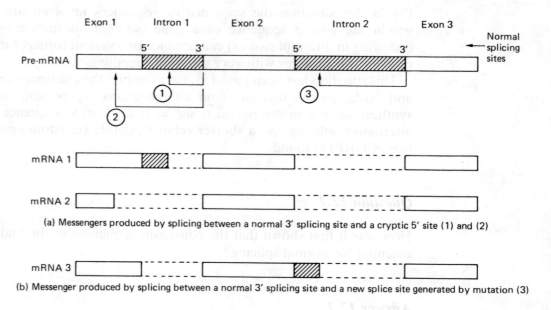

(a) Messengers produced by splicing between a normal 3' splicing site and a cryptic 5' site (1) and (2)

(b) Messenger produced by splicing between a normal 3' splicing site and a new splice site generated by mutation (3)

Figure 12.13 *Mutations affecting splicing in the human β-globin gene. The dotted lines represent the sequences deleted from the messengers as a result of altered splicing. The final result is the production of abnormal polypeptides*

These experiments clearly show the key role that consensus sequences play in splicing.

NOTE

These experiments were reported by R. Treisman, S. H. Orkin and T. Maniatis in 1983.

Question 12.8

What are enhancers and how do they differ from most other regulatory sequences?

Answer 12.8

Enhancers are sequences which stimulate the transcription of adjacent structural genes. They are activated by the binding of a specific protein and then (so it is thought) act as sites for the assembly of transcriptional initiation complexes.
They differ from most other control sequences as follows:

(1) They may be several kilobases away from the gene whose transcription they control.
(2) They may be located either upstream or downstream of the gene.

(3) They act in either orientation and so can simultaneously influence the expression of two genes, one on each side of the enhancer sequence.

(4) They must be located on the same molecule of DNA as the regulated gene, but the sequence can be on either DNA strand.

(5) They are not gene-specific; an enhancer will stimulate any gene placed beside it.

(6) Enhancers are, however, tissue-specific. Thus, the enhancer for the immunoglobulin gene will only stimulate transcription of an adjacent gene in the cells of the immune system.

(7) Enhancers preferentially stimulate transcription from the nearest promoter.

(8) Hormone receptor proteins are among the proteins that bind to enhancers; thus, enhancers may play an important role in regulating gene activity during development.

Question 12.9

How do upstream activating sequences control expression of the *gal* genes in yeast?

Answer 12.9

In yeast there are three co-ordinately regulated genes whose products catalyse successive steps in the conversion of galactose to glucose-6-phosphate and, although closely linked, they are separately transcribed. In this system (Figure 12.14) there are two UASs, one in the region between *gal1* and *gal10* which also contains the promoter sequences for these genes and the other adjacent to the promoter for *gal7*. There are two unlinked regulatory genes: (1) a positive regulatory gene producing an activator protein which binds to both UASs and stimulates transcription of each of the *gal* genes and (2) a negative regulatory gene encoding a repressing protein. In the absence of galactose, transcription is not stimulated as, probably, the repressing protein binds to and inhibits the

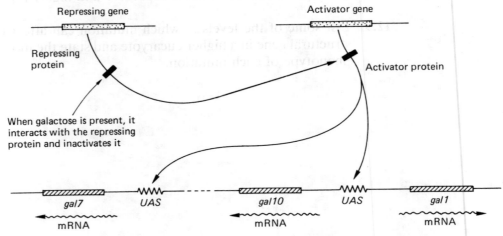

Figure 12.14 *Regulation of the* gal *genes in yeast*

activator protein; however, if galactose is present, transcription is stimulated, as galactose appears to interact with the repressing protein, destroying its capacity to inhibit the activator protein.

NOTES

1 *gal1*, *gal7* and *gal10* encode galactokinase, galactose transferase and galactose epimerase, respectively.
2 If any other yeast gene is placed near this UAS, then its activity is controlled by the presence or absence of galactose.

12.9 Supplementary Questions

12.1 What species of eucaryotic RNA does not require processing?

12.2 RNA polymerase III specifically transcribes small molecules of RNA. Why does it not transcribe the 5.8 S rRNA?

12.3 You have succeeded in isolating a mutation in the promoter sequence of a tRNA gene. What effect would you expect this to have on (**a**) the gene product and (**b**) the phenotype of the cell or organism?

12.4 What are Housekeeping Genes?

12.5 Why is splicing a very accurate process?

12.6 What would be the outcome if splicing occurred between the GU sequence of one intron and the AG sequence of an adjacent intron?

12.7 Do you consider splicing to have any real advantage over the simpler situation found almost universally in procaryotic systems?

12.8 Distinguish between middle-repetitive and highly repetitive DNA.

12.9 List some of the levels at which mutation can affect the expression of a structural gene in a higher eucaryote and state the most obvious molecular phenotype of each mutation.

13 Recombinant DNA technology

13.1 Introduction

Until the advent of gene cloning in the early 1970s, it was almost impossible to study individual genes, particularly eucaryotic genes, at a molecular level, as the technology for producing the required amounts of pure genic DNA did not exist. To some extent bacterial geneticists had overcome this problem by linking certain bacterial genes to phage genomes (by using transducing phages), but no similar method was available to eucaryotic geneticists. Thus, until the advent of gene cloning, a technique which itself was only made possible by the discovery of restriction endonucleases, little was known about the molecular structure and the functioning of eucaryotic genes; recombinant DNA technology and gene cloning overcome this problem by linking eucaryotic (or procaryotic) genes to plasmid or phage vectors which can be easily replicated in a bacterial host.

In this chapter we shall consider only the most fundamental aspects of these important technologies.

13.2 An Outline of Gene Cloning

A simple cloning experiment involves the following stages:

(1) A linear fragment of DNA carrying the gene to be cloned is isolated or synthesised *in vitro* and, using recombinant DNA technology (Sections 13.3 and 13.4), inserted into a second DNA molecule called a **cloning vector** (Section 13.5) or **vector** (Figure 13.1). This vector is usually a circular molecule and is commonly either a small multicopy plasmid or phage. The vector is genetically marked so as to facilitate the manipulation and recognition of the new recombinant DNA molecule.

(2) The recombinant vector DNA is introduced into a host cell in which it can replicate along with the endogenous DNA. Frequently *E. coli* is the host and the vector is introduced into it by transformation.

(3) The host cells are allowed to replicate, forming a **clone** of genetically identical cells, each cell containing from one to many copies of the recombinant DNA molecule.

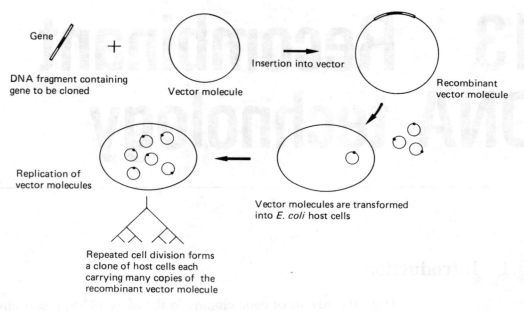

Figure 13.1 *A simple cloning experiment*

(4) The cloned recombinant DNA can be isolated from the host cells, purified and studied as required. It can, for example, be sequenced or it can be transcribed and translated and the gene product isolated and characterised, and the regulation of gene expression can be studied in isolation from the rest of the genome.

13.3 Restriction Endonucleases

When a bacterial cell is invaded by viral DNA (or by any other 'foreign' DNA), the viral infection may be **restricted**, because the invading nucleic acid is attacked by an enzyme called a **restriction endonuclease** which cuts it at one or more specific sites.

Very many different restriction endonucleases have been isolated from a wide variety of bacterial species and all act by attacking DNA containing particular nucleotide sequences, 4–8 bp long. In recombinant DNA technology the most important restriction endonucleases are the type II enzymes, which *always* cut the DNA between *particular* nucleotides within the recognition sequence or **restriction site** (the type I enzymes also cut the DNA but at some distance from the recognition sequence).

Although each bacterial strain has at least one restriction endonuclease, these enzymes do not attack the DNA of their own cells. This is because each strain has not only a specific restriction endonuclease, but also a complementary **modification enzyme** or **methylase** which recognises the same target sequence and, each time it occurs in the host DNA, methylates particular bases within it.

One widely used restriction endonuclease is *Eco*R1, isolated from a strain of *E. coli*. This recognises the palindromic sequence (most restriction sites are palindromes)

```
5'  G   A   A   T   T   C
3'  C   T   T   A   A   G
```

and cuts (arrows) the DNA between the G and A in every GAATTC sequence, generating base-complementary sticky or cohesive ends (see Question 3.7):

```
5' — — — G                    A  A  T  T  C  — — — 3'
                     and
3' — — — C  T  T  A  A                 G  — — — 5'
```

Most *E. coli* strains contain a methylase which recognises the 5′ AATTC 3′ sequences and methylates one of the adenine residues protecting the *E. coli* DNA from attack by *Eco*R1 but not from attack by other restriction endonucleases.

Other restriction endonucleases recognise different target sequences and generate either different sticky ends or blunt ends. *Bam*H1 (from *Bacilus amyloliquifaciens*) recognises the sequence 5′ GGATCC 3′ and creates different sticky ends

```
5' — — — G                    G  A  T  C  C  — — — 3'
                     and
3' — — — C  C  T  A  G                 G  — — — 5'
```

while *Hae*III (from *Haemophilus aegyptius*) recognises the sequence 5′ GGCC 3′ and generates the blunt ends

```
5' — — — — G  G           C  C  — — — — 3'
                     and
3' — — — — C  C           G  G  — — — — 5'
```

The discovery of these enzymes in the late 1960s paved the way for the development of recombinant DNA technology.

13.4   Producing Recombinant DNA Molecules

Suppose that we wish to clone a gene by inserting it into a plasmid known to have a single *Eco*R1 restriction site. First, the fragment of DNA containing the gene to be cloned is produced by treating the donor DNA with *Eco*R1; this is followed by treating the plasmid with *Eco*R1 (cutting the plasmid DNA at the single *Eco*R1 restriction site) and mixing together the two populations of DNA molecules. Since *Eco*R1 cleaves DNA within each copy of a specific palindromic sequence, both the donor DNA and the cleaved plasmid DNA will *always* have *identical* 'sticky' ends (Figure 13.2). Thus, the donor fragment can be annealed into the plasmid DNA by complementary base pairing and the remaining single-strand gaps sealed with DNA ligase, producing a molecule of **recombinant DNA**.

This is a highly specific process and, in general, two molecules with different sticky ends (that is, molecules produced by the activity of different restriction endonucleases) cannot be ligated together.

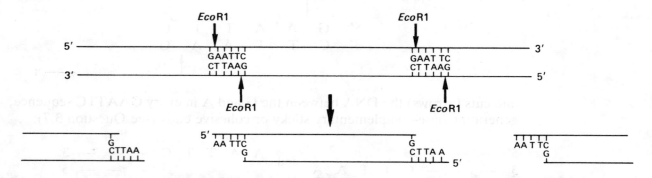

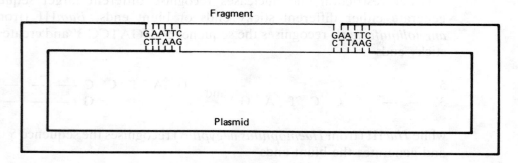

(A) *Eco*R1 cleaves DNA at every 5′ GAATTC 3′ sequence producing fragments of DNA with single-stranded cohesive ends

(B) If plasmid DNA with a single *Eco* R1 restriction site is also cleaved with *Eco* R1, the fragment can be inserted into the plasmid DNA and the remaining single-strand breaks sealed with ligase

Figure 13.2 *A recombinant DNA molecule synthesised using EcoR1*

It is also possible to join together two blunt-ended molecules, but this is a relatively inefficient and non-specific process.

13.5 Cloning Vectors

The vector is always an independent replicon, so that the recombinant DNA molecule can multiply when introduced into a host cell. There are very many different cloning vectors but one commonly used plasmid vector is pBR322. This plasmid (Figure 13.3) is only 4363 bp in size, so that it can be easily manipulated and purified, and it carries two genes for antibiotic resistance — *amp*ʳ (ampicillin resistance) and *tet*ʳ (tetracyline resistance). Each of these genes contains a unique restriction site, so that a gene to be cloned can be inserted into either *amp*ʳ or *tet*ʳ; this destroys the integrity of the gene (it is called **insertional inactivation**), so that the recombinant plasmid can only confer resistance to the other antibiotic. This dual drug resistance is an extremely valuable property, as normal *E. coli* cells are sensitive to both drugs; host cells carrying pBR322

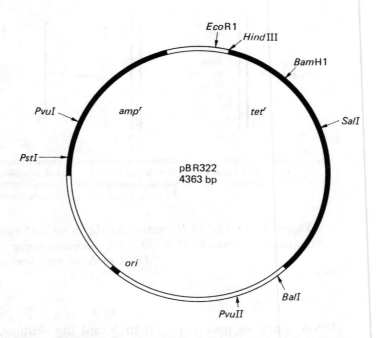

Figure 13.3 *Plasmid pBR322. The figure shows the genetic map and the location of some of the restriction sites that occur only once. For other restriction endonucleases there may be more than one restriction site; for example, there are 10 or more* AluI *restriction sites*

without a cloned gene are resistant to both ampicillin and tetramycin, while any host cells carrying pBR322 where the gene to be cloned has been inserted into *amp*^r will only be resistant to tetracycline.

Another feature of pBR322 is that it is a multicopy plasmid and there are normally about 15 copies per cell, but this can be increased to 1000–3000 copies by growing the host cells in the presence of chloramphenicol. Thus, it is possible to isolate relatively large amounts of plasmid DNA containing the cloned gene.

13.6 Cloning DNA *in vitro*

Very recently (1988) a new technique, the **polymerase chain reaction** (PCR), has been developed and appears to be revolutionising cloning protocols. This technique allows the *in vitro* amplification of specific nucleotide sequences, the only pre-requirement being that parts of the nucleotide sequences flanking the region to be amplified be known. The method, in outline, is as follows (Figure 13.4):

(1) Two DNA oligonucleotides are chemically synthesised; these are 15–20 nucleotides long and each is base-complementary to one of the flanking sequences on the opposite strands of the DNA double helix.

(2) The fragment of chromosomal DNA to be cloned is denatured and the oligonucleotides are annealed with the separated strands.

(3) The annealed mixture is incubated in the presence of DNA polymerase and the four nucleoside triphosphates; the oligonucleotides act as primers for

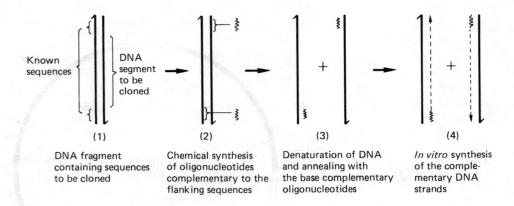

(1)	(2)	(3)	(4)
DNA fragment containing sequences to be cloned	Chemical synthesis of oligonucleotides complementary to the flanking sequences	Denaturation of DNA and annealing with the base complementary oligonucleotides	*In vitro* synthesis of the complementary DNA strands

Figure 13.4 *The PCR method for the* in vitro *cloning of DNA. The process is automated to repeat itself 20–30 times, each successive cycle doubling the number of DNA molecules present*

DNA synthesis and in less than 5 min the sequences between the two oligonucleotides have been copied.

(4) The process is automatically repeated 20–30 times, so that within a few hours many millions of double-stranded DNA copies have been produced.

13.7 Applications of Recombinant DNA Technology

The development of recombinant DNA technology has produced one of the most dramatic transitions in the history of genetics; although the methods were originally developed for pure research, it was soon realised that they had enormous commercial potential in the fields of medicine, industry, agriculture and plant breeding. Only a few of the many diverse applications will be indicated here.

(a) Applications in Pure Research

Without doubt the most important recombinant DNA technique is **DNA sequencing** (establishing the order of the nucleotides along molecules of DNA), as this has provided much of our detailed knowledge of gene structure and function; many examples of nucleotide sequences and their functions are given in this book.

Efficient methods for sequencing DNA were developed independently by Allan Maxam and Walter Gilbert, working at Harvard (the chemical cleaveage method) and by Fred Sanger at Cambridge (the chain termination method). Both of these techniques require the use of restriction enzymes and cloning of the DNA fragment to be sequenced, but a description of them is outside the scope of this book.

(b) Applications in Medicine

Hitherto many important biological compounds could only be isolated from cells or chemically synthesised in very small amounts; these compounds include somatostatin (Question 13.5), α-interferon (an antiviral agent) and insulin (used in the treatment of diabetes). Genes for these compounds have now been isolated and cloned into bacteria which can then produce these compounds in large amounts. This not only increases their availability, but also reduces their cost. Cloning is also being used to produce certain synthetic vaccines.

(c) Applications in Industry

Industry uses very large amounts of microbial enzymes; these, and many industrially important chemicals, are now produced more efficiently and at a lower cost by using genetically engineered micro-organisms.

An interesting development has been the production of a plasmid containing genes derived from several species of bacteria and capable of metabolising petroleum; this plasmid, when present in a marine bacterium, can assist in the breakdown of oil spillages.

(d) Applications in Plant Breeding and Agriculture

Genetic engineering can speed up the development of new strains by overcoming the species barrier for gene transfer. For example, in plant breeding it is extremely laborious to transfer a particular gene for disease resistance into a hybrid strain by the use of conventional genetic crosses but this is relatively easy to achieve by genetic engineering; furthermore, it enables the transfer of a gene from one species to a quite different species, something that the conventional plant breeder is unable to do.

Glyphosate is an important component of many herbicides and acts by inhibiting an essential enzyme present in many plant species, but these herbicides cannot generally be used as selective weedkillers because they kill both the crop and the weeds. However, a mutant strain of the bacterium *Salmonella typhimurium* carries a gene conferring resistance to glyphosate; this gene has been cloned onto a plasmid obtained from *Agrobacterium* (a bacterium causing gall tumours on plants) and infecting plants of maize, cotton or tobacco with this plasmid produces glyphosate-resistant strains, enabling glyphosate to be used as a selective weedkiller.

13.8   Questions and Answers

Question 13.1

Use an annotated diagram to show how you could most simply clone a gene using pBR322 as a vector.

Answer 13.1

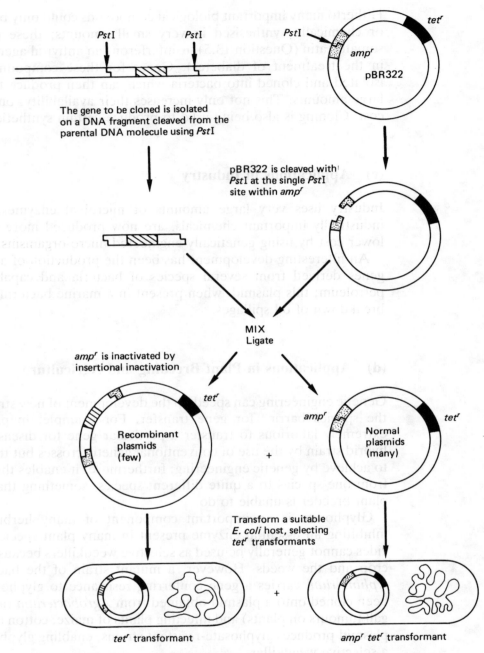

The gene to be cloned is isolated on a DNA fragment cleaved from the parental DNA molecule using *Pst*I

pBR322 is cleaved with *Pst*I at the single *Pst*I site within *amp*^r

MIX
Ligate

amp^r is inactivated by insertional inactivation

Recombinant plasmids (few)

+

Normal plasmids (many)

Transform a suitable *E. coli* host, selecting *tet*^r transformants

tet^r transformant

+

amp^r *tet*^r transformant

The two classes of transformant can be distinguished by replating on medium containing ampicillin

Figure 13.5 *A simple cloning experiment using pBR322*

NOTE

Host cells transformed by pBR322 are isolated by selecting *tet*^r colonies on minimal medium containing tetracycline. The ampicillin-sensitive and tetracycline-resistant clones (carrying the recombinant plasmids) are distinguished from the clones resistant to both ampicillin and tetracycline (carrying non-recombinant plasmids) by replating individual *tet*^r colonies on medium containing ampicillin; those that do NOT grow when replated have had the *amp*^r gene inactivated by insertional inactivation.

Question 13.2

How does cloning enable fragments of DNA containing individual genes to be purified?

Answer 13.2

Large molecules of DNA may have many restriction sites for any particular restriction endonuclease, so that treatment of a whole chromosome or a whole genome with a restriction enzyme will produce many different DNA fragments, each carrying a different gene or a different small segment of the genome; only one of these fragments carries the gene to be cloned.

When this population of fragments is mixed with vector molecules treated with the same restriction endonuclease (Figure 13.6), each fragment will be inserted into a different vector molecule (say a plasmid), and likewise, when the host cells are transformed with this population of plasmid molecules, each host cell will only take up a single molecule of plasmid DNA and each colony selected as containing a recombinant plasmid will, in fact, carry multiple copies of just one specific fragment of donor DNA. Once the clone carrying the gene to be studied has been identified, the recombinant DNA molecules can be extracted from the host cells and purified.

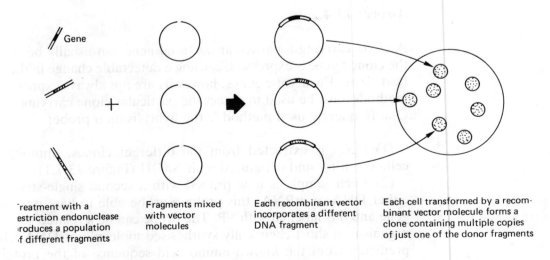

| Treatment with a restriction endonuclease produces a population of different fragments | Fragments mixed with vector molecules | Each recombinant vector incorporates a different DNA fragment | Each cell transformed by a recombinant vector molecule forms a clone containing multiple copies of just one of the donor fragments |

Figure 13.6 *The purification of DNA fragments by gene cloning*

Question 13.3

What is a gene library?

Answer 13.3

A gene library, or colony bank, is a set of bacterial clones, each clone harbouring a plasmid or phage vector containing a different DNA fragment from a particular donor species; the collection is of such a size that there is a 95–99% probability of every segment of genomic DNA being present in at least one clone. Once a gene library has been assembled, any particular clone can be withdrawn and the genes it carries can be studied, thus avoiding the time-consuming process of going through the cloning process whenever it is required to clone a particular gene.

In order to reduce the number of clones in the collection, the most commonly used vector is a derivative of phage lambda which can accommodate between 12 and 20 kb of donor DNA — this is in contrast to pBR322, where the maximum size of the inserted DNA is about 5 kb. Using a lambda vector, the library for *Escherichia coli* consists of some 400 clones and these contain every (or most) *E. coli* gene(s); the equivalent library for the human genome would contain about 250 000 clones.

Gene libraries are now available for many organisms, including *E. coli*, yeast, *Drosophila melanogaster* and man.

Question 13.4

You have prepared a gene library; how might you now identify the particular clone carrying a gene of interest?

Answer 13.4

A clone carrying a particular bacterial gene can usually be recognised because the cloned gene is expressed, causing a detectable change in the phenotype of the host clone. Eucaryotic genes, however, are not always expressed and alternative methods must be used to detect the particular clone carrying the required gene. One frequently used method is the hybridisation probe:

(1) DNA is extracted from the different clones, immobilised on a nitro-cellulose filter and denatured with NaOH (Figure 13.7.1).

(2) Each sample is now probed with a second single-stranded nucleic acid, either DNA or mRNA; this probe *must* be able to base pair with the required gene and it is labelled with ^{32}P. The probe can be a related gene from a different organism, a short chemically synthesised molecule of DNA whose sequence is predicted from the known amino acid sequence of the protein or messenger RNA specific to the gene (Figure 13.7.2).

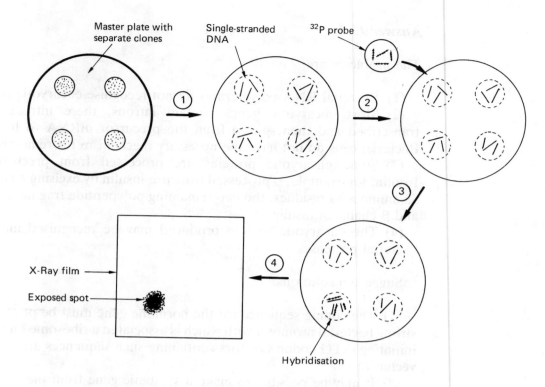

Figure 13.7 *An* in situ *hybridisation assay.*
(1) DNA is extracted from each clone, denatured and fixed to a nitrocellulose filter.
(2) Each spot of DNA is probed with ^{32}P*-labelled base-complementary single-stranded DNA or RNA.*
(3) The probe only hybridises with the DNA that includes the required gene; washing the filter removes any free (non-hybridsed) probe.
(4) The filter is overlaid with X-ray film; after exposure and development a black spot shows which sample contains the required gene. Finally, the required clone is recovered from the master plate

(3) The probe will base pair (hybridise) only with the gene of interest, and when the filter paper is washed, any unbound labelled probe will be removed (Figure 13.7.3).

(4) The filter is dried and overlaid with X-ray film. The emission from the ^{32}P label exposes the film and, after development, the required clone is indicated by a black spot on the film.

This is known as the **colony** or *in situ* **hybridisation assay**.

Question 13.5

You are considering using *E. coli* as a host in which to clone a gene encoding an animal hormone with the ultimate object of producing this hormone in the host cells. Summarise the major problems you might encounter in obtaining expression of the cloned gene and suggest how you might overcome these.

Answer 13.5

The problems are:

(1) Bacterial RNA polymerases do not recognise eucaryotic promoters.

(2) Most eucaryotic genes contain introns; these introns are normally transcribed and then spliced from the precursor mRNA to form messenger. Bacterial cells do not have the necessary mechanism for removing introns.

(3) Some eucaryotic proteins are processed from precursor molecules. Insulin, for example, is processed from pre-insulin by excising an internal tract of 33 amino acid residues, the two remaining polypeptide fragments forming the α and β chains of insulin.

(4) The eucaryotic protein produced may be recognised and degraded by bacterial proteases.

Suggested solutions:

(1) The coding sequence for the hormone gene must be placed adjacent to a strong bacterial promoter with which is associated a ribosome binding site and an initiating ATG codon (vectors containing such sequences are called **expression vectors**).

(2) It may be possible to make a synthetic gene from the hormone mRNA, using the enzyme reverse transcriptase which makes DNA from an RNA template. This DNA will not contain introns and it can be inserted into the vector and cloned. Alternatively, if the protein is very short, a gene can be chemically synthesised; this synthetic gene would contain an ATG initiation codon, the coding sequence for the gene predicted from the known amino acid sequence of the hormone, and one or two stop codons:

$$\text{ATG} \text{------- coding sequence -------} \text{TGA TAG}$$

This synthetic gene can now be inserted into the vector.

(3) In some instances processing can be carried out *in vitro*. Alternatively, if processing is a problem, it may be possible to use a synthetic gene, which does away with the need for processing.

(4) It may be possible to eliminate the protease from the host cells by isolating a suitable mutant.

Many of these problems would be more easily solved by using yeast as a host, especially since there are now plasmids, called **shuttle vectors,** which are capable of replicating in both *E. coli* and yeast.

NOTES

1 ATG, TGA and TAG are the DNA sequences corresponding to AUG, UGA and UAG, translational start and stop signals in the mRNA.

2 The chemical synthesis of a gene was the method used in cloning the somatostatin gene. Somatostatin is a hormone produced by the hypothalamus and it is only 14 amino acids long, so that the synthetic gene, including the

translational start and stop signals required for expression in the bacterial host, was only 51 bp long.

13.9 Supplementary Questions

13.1 What is complementary DNA (cDNA)?

13.2 The chromosome of phage lambda is 48.6 kb long and contains 5 *Eco*R1 restriction sites (5'GAATCC3') and 28 *Pst*I sites (5'CTGCAG3') sites. How many such sites would you expect to be present on a molecule of this size? What does the observation suggest?

13.3 The sequence 5' CGAACATATGGAGT 3' contains a 6 bp recognition sequence for a type II restriction endonuclease. What is the probable sequence of the recognition site?

13.4 Suggest how a fragment of DNA with blunt ends might be inserted into an *Eco*R1 restriction site.

13.5 A plasmid carrying genes for resistance to ampicillin and kanamycin is treated with *Eco*R1, which cuts uniquely within the kanamycin gene. The digest is then ligated with *Eco*R1-treated yeast DNA and used to transform a strain of *E. coli* sensitive to both drugs.

  (a) What pattern of drug resistance should you select so as to recover cells that have received a plasmid?
  (b) How would you distinguish those clones that have received a plasmid containing a yeast DNA insert?

13.6 What is a shuttle vector?

Answers to supplementary questions

Chapter 1

1.1 Meiosis halves the chromosome number during gamete formation, so that, when two haploid gametes fuse to form a zygote, the diploid number of chromosomes is restored.

1.2 **Centrioles** are organelles found only in animal cells. At the onset of cell division two daughter centrioles move to opposite 'ends' of the cell, where they organise the assembly of the microtubules that form the spindles; they form the poles of the spindles that form during cell division.
Centromeres are specialised late-replicating regions of the chromosomes to which the spindle fibres attach during cell division. They control the movement of the chromosomes.

1.3 In different species the chromosomes are, at the most, only partly homologous and the nucleotide sequences along the molecules of DNA are quite different. As a consequence, the chromosomes are unable to form homologous pairs at meiosis and so cannot correctly orientate on the metaphase plate.

1.4 B is $2n + 1$ and has two X chromosomes and a Y chromosome. It is from a fertile female. C has 2 X chromosomes and is from a female. However, the tip of one chromosome II is missing (it has been **deleted**) and the fly probably showed some abnormal characteristics. D is XO ($2n - 1$) and, therefore, from a sterile male.

Chapter 2

2.1 (a) 5' CCTGGCATGAT 3'; (b) 5' CCUGGCAUGAU 3'.

2.2 It is a trinucleotide; all the other nucleotides are mononucleotides, as they lost two phosphate groups during phosphodiester bond formation.

2.3 Determine the base ratios. If thymine is present, it is DNA; if uracil, it is RNA. If the molecule is double-stranded, there will be equivalent amounts of A and T (or U) and of G and C.

2.4 DNase degrades DNA and RNase degrades RNA.
Exonucleases attack a nucleic acid strand from one or both ends.
Endonucleases make cuts within a nucleic acid strand.

2.5

Parental duplex	Daughter DNA duplices when replication is		
	Conservative	Semi-conservative	Dispersive
‖	‖ + ‖	‖ + ‖	(dispersive figure) + (dispersive figure)

———— Parental DNA ————Newly replicated DNA

2.6 The DNA is double-stranded and is replicating. The segments of RNA are the short primer sequences involved in lagging strand synthesis.

2.7 In DNA replication ligase is essential for lagging strand synthesis, as it makes the bond between the 5′ end of each Okazaki fragment and the 3′ end of each preceding fragment (see Figure 2.5). RNA synthesis occurs along either a DNA (RNA polymerase activity) or an RNA (RNA replicase activity) template strand; this corresponds to leading strand synthesis and is always continuous in the 5′ to 3′ direction, so that ligase is not required.

2.8 After one generation half of the DNA molecules would be 'heavy' and half would be 'light'. After two generations one-quarter would be 'heavy' and three-quarters 'light':

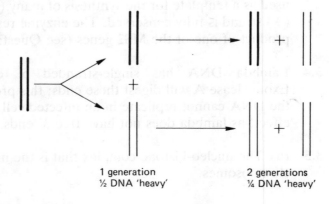

1 generation
½ DNA 'heavy'

2 generations
¼ DNA 'heavy'

3.1 (a) Replication from a single origin

(b) Replication from two origins.

In this experiment it is not possible to determine whether replication is unidirectional or bidirectional.

3.2 The cells are grown synchronously in **low**-activity tritiated thymidine and just before the end of a replication cycle fed with **high**-activity tritiated thymidine. The cells are allowed to enter the next replication cycle and the DNA is then extracted and autoradiographed. With unidirectional replication the terminus and the origin of replication will coincide; there will be a single densely labelled region and a single replication fork (A). If replication is bidirectional, the terminus and the origin will be separate and at opposite sides of the molecule; there will be two densely labelled regions and two replication forks (B):

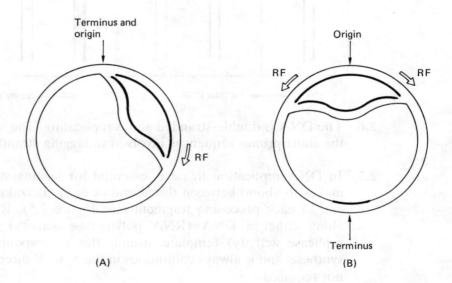

3.3 Phage MS2 has a single strand of RNA as its chromosome and this (+) strand replicates as follows: (1) the (+) strand acts as a template for the synthesis of a base-complementary (−) strand and (2) this (−) strand is used as a template for the synthesis of many (+) viral strands. The original (+) strand is fully conserved. The enzyme responsible is RNA replicase, a product of one of the MS2 genes (see Question 10.10).

3.4 Lambda DNA has single-stranded 5′ ends (the cohesive ends). Exonuclease A will digest these ends; this prevents circularisation, so that the DNA cannot replicate in an infected cell. Exonuclease B will have no effect, as lambda does not have free 3′ ends.

3.5 (a) The nucleo-histone complex that is the main component of eucaryotic chromosomes.

(b) The structures carrying the genetic information; a molecule of nucleic acid in viruses and procaryotes, a complex of nucleic acid and protein in eucaryotes.

(c) A daughter chromosome is called a chromatid until its centromere divides.

(d) A laterally differentiated region of a eucaryotic chromosome. *Either* one of the bead-like structures seen at meiotic prophase *or* a band of a polytene chromosome.

3.6 During the preceding replication two sister chromatids have broken (or recombined) at homologous positions and reunited in a new combination:

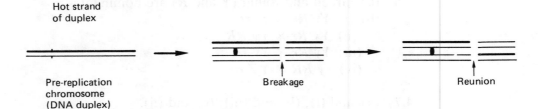

Hot strand of duplex

Pre-replication chromosome (DNA duplex)

Breakage

Reunion

This is called **sister chromatid exchange** (SCE). The frequency of these events is considerably increased if the cells are irradiated or treated with a chemical mutagen (Chapter 11).

3.7 (1) It is permanently condensed and dark-staining; euchromatin is only condensed during cell division. (2) Relative to euchromatin it is over-condensed at meiotic prophase and undercondensed at metaphase. (3) It is found in the centromeric regions. (4) In many cells blocks of heterochromatin tend to associate with the inner surface of the nuclear membrane. (5) In salivary gland cells the blocks of constitutive heterochromatin associate to form the chromocentre. (6) It is later-replicating than euchromatin. (7) It contains few — if any — functional genes and it is doubtful whether it is ever transcribed.

3.8 It shows that the DNA is organised into regularly repeating units about 200 bp long and that only one small region within each repeating unit is susceptible to nuclease attack. In terms of the nucleosome model, the DNA associated with the histone octamer is protected from nuclease attack and the DNA can only be cut within the tracts of linker DNA.

Chapter 4

4.1 In a cross between two pure-breeding strains differing by a single character, the phenotype or allele expressed in the F1 progeny is *dominant* and the unexpressed character or allele is *recessive*.

4.2 If, for example, a vestigial-winged female of *D. melanogaster* is crossed with an ebony-bodied male, then the reciprocal cross is between an ebony female and a vestigial male.

4.3 (1) Establish pure breeding lines of white and purple flowered plants.
(2) Intercross the two lines.
(3) Examine the F1 phenotypes (determines which allele is dominant).
(4) Self-cross or intercross the F1 progeny. A 3:1 ratio demonstrates inheritance due to a single gene difference.

4.4 The parents were homozygous for *different* recessive genes, i.e *AAbb* × *aaBB*, so that all the children would be *AaBb* double heterozygotes.

4.5 Chance of first child being albino is 1 in 8. Chance of both members of a pair of non-identical twins being albino is 1 in 64.

4.6 (a) Green and round (*Y* and *R*) are dominant.
  (b) (i) *Yy Rr* × *yy rr*
    (ii) *Yy Rr* × *Yy RR*
    (iii) *Yy Rr* × *Yy rr*
    (iv) *yy RR* × *YY rr*

4.7 (a) and (i); (b) and (iii); (c) and (ii).

Chapter 5

5.1 (a) The tendency for genes on the same chromosome to remain together when they enter the gametes, so that the parental character combinations are more frequent among the progeny than are the non-parental combinations.

(b) Because crossing over can result in the exchange of genetic material between two homologous non-sister chromatids.

5.2 Because each crossover involves only two of the four chromatids — the other two chromatids always remain non-recombinant.

5.3 Make the corresponding *coupling* backcross $a^+b/ab^+ \times ab/ab$. If linkage is the explanation, then the parental types are now a^+b/ab and ab^+/ab and these will make up 80% of the progeny. If differential viability is the explanation, then the same results will be found in both crosses.

5.4 (a) Genes located on the same chromosomes and so showing evidence of linkage to other genes in the same linkage group.

(b) Four.

5.5 In yeast and certain other micro-organisms there is no morphological differentiation into males and females, but only haploid cells of different mating type can fuse to form a diploid zygote. In yeast these physiologically differentiated cell types, the **mating types**, are known as *a* and α; when these mate, they form an *a*/α diploid.

5.6 There must be a crossover between the locus under investigation and its centromere.

5.7 Centromeres *always* segregate at the first meiotic division, and the closer a gene is to its centromere the greater is the proportion of first division segregation asci. Thus, to measure the frequency of second division segregation between an unmapped gene and its centromere, you must carry out a two-factor cross between the unmapped gene and a gene known to be very closely linked to its centromere — in effect, this 'labels' the centromere. Each crossover between the two loci corresponds to a crossover between the unmapped locus and the centromere. This produces tetratype asci.

5.8 The markers *must* be in repulsion, i.e. *sn* +/+ *y*, and there has been a crossover between the *sn* locus and the centromere:

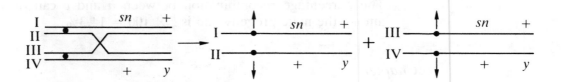

At anaphase chromosome I can segregate with either chromosome III or chromosome IV. In the former instance the result will be a twin spot with adjacent patches of yellow (II + IV, + *y*/+ *y*) and singed (I + III, *sn* +/*sn* +).

5.9 Mitotic recombination occurs between homologous but non-sister chromatids; sister chromatid exchange occurs between two sister chromatids.

5.10 (a) The maps are compiled by measuring the percentage recombination values between linked markers in many different experiments and summating the results. Thus, the map distance between *al* and *c* (75.5 r.u.) is the sum of the map distances between *al* and *dp* (13 r.u.), between *dp* and *pr* (41.5 r.u.) and between *pr* and *c* (21 r.u.).
(b) Any marker pairs separated by 50 or more recombination units will assort independently.

5.11

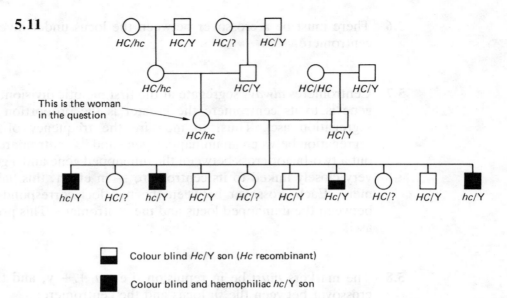

This is the woman in the question

■ Colour blind *Hc*/Y son (*Hc* recombinant)

■ Colour blind and haemophiliac *hc*/Y son

? The remaining alleles present cannot be deduced from the information provided

The percentage recombination between *h* and *c* can only be estimated among the male progeny and is $\frac{1}{7} \times 100 = 14.3\%$.

Chapter 6

6.1 Transfer replication commences from a site within the integrated F factor, so that only part of F constitutes the leading end of the transferred chromosome; the remainder of F is only transferred in the exceptional instances where the entire bacterial chromosome is transferred.

6.2 Different Hfr strains have F integrated at many different sites around the bacterial chromosome and in either orientation; it is the location and orientation of F that determines the origin and the direction of transfer.

6.3 Use an Hfr where F is integrated beside *gal*+ and where *gal*+ is one of the very last markers to be transferred (such as Hfr2 in Figure 6.8). Make an F⁻ *gal*⁻ × Hfr *gal*+ cross, interrupting mating after 30 min and selecting *gal*+ exconjugants. Under these conditions the only *gal*+ exconjugants will have received an F′*gal*+ plasmid from a rare F′*gal*+ bacterium that has arisen spontaneously in the population of Hfr cells.

6.4 These plates are controls. The first control detects and provides an estimate of the frequency of spontaneous mutations to the wild type; the second control verifies the sterility of the phage suspension (i.e. shows it does not contain any surviving donor bacteria).

6.5 Isolate the F and F′ plasmids and gently heat the DNA so that it separates into its component strands. The denatured DNAs are mixed and cooled gently, so allowing the DNA to reanneal. Some of the duplex molecules will be heteroduplexes made up of one strand from F and one from F′.

Regions of non-homology are seen under the electron microscope as single-stranded loops:

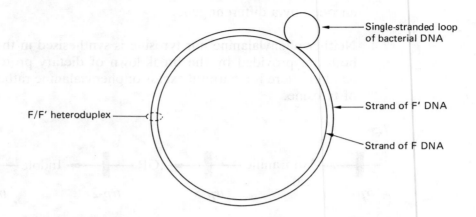

- Single-stranded loop of bacterial DNA
- Strand of F' DNA
- Strand of F DNA

F/F' heteroduplex

6.6 It may (1) persist without replicating and allow the formation of an abortive transductant, (2) be recombined into the bacterial chromosome and form a transductant, or (3) be degraded by nucleases.

6.7 a^+ and b^+ are too far apart to be co-transformed. If they were co-transformable, there would have been many more a^+b^+ co-transformants when a^+b^+ donor DNA was used than when a mixture of a^+b^- and a^-b^+ DNAs was used.

6.8 (**a**) Hfr transfer and interrupted mating; (**b**) transduction.

6.9 *rII* mutants will not grow on *E. coli* strain K; however, if one mutant is *rIIA* and the other *rIIB*, complementation will occur and phage growth will proceed normally. If recombination were involved, some wild type (*rII*$^+$) recombinants would be present among the phage progeny, but with complementation only the original *rIIA* and *rIIB* mutants will be recovered.

6.10 P1 packages about 100 kb of DNA (2.5% of the *E. coli* chromosome); in transformation only markers less than 10 kb (0.25%) can be cotransformed. Hence, (**a**) neither pair of markers is co-transformable and (**b**) only *C* and *D* can be co-transduced.

Chapter 7

7.1 (**a**) Its three-dimensional structure; (**b**) its amino acid sequence.

7.2 If the substitution occurs in a functionally important part of the polypeptide, then its three-dimensional configuration, and, hence, its activity, will be altered. Otherwise the substitution may not have any noticeable effect.

7.3 (1) Not all genes encode proteins — some encode tRNA and rRNA.

(2) Not all proteins are enzymes — but they are all gene-encoded.

(3) Some proteins are assembled from two or more different proteins, each encoded by a different gene.

7.4 Neither phenylalanine nor tyrosine is synthesised in the human body and both are provided by the breakdown of dietary protein. Thus, in PKU children there is a harmful excess of phenylalanine rather than a deficiency of tyrosine.

7.5

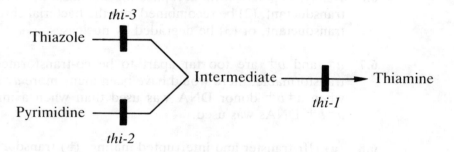

7.6 The unusual pattern of response suggests a branched pathway:

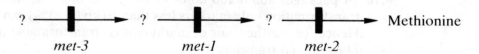

7.7 Bacteriophages: simultaneous infection with two mutants

Bacteria: abortive transduction

                       making F′ partial diploids

Eucaryotes: heterocaryon formation (ascomycetes)

                       constructing doubly heterozygous strains

7.8 The mutants are blocked at successive stages of methionine synthesis:

$$? \longrightarrow ? \longrightarrow ? \longrightarrow \text{Methionine}$$

met-3   met-1   met-2

7.9 **(a)** Only found where the active protein is a dimer of two identical polypeptide sub-units.

(b) Only occurs when the two mutant polypeptides are altered in precisely complementary ways so that the two altered sub-units can dimerise to form an active protein.

7.10 Consider sickle-cell anaemia and assume that the term is used to qualify the allele. Haemoglobin S is present in both Hb^S/Hb^S and Hb^A/Hb^S individuals so that the allele can be called dominant; however, the anaemia is only expressed in Hb^S/Hb^S individuals and so can be called recessive! These difficulties do not arise if the terms 'dominant' and 'recessive' are used to describe the characters (this is how these terms were used by Mendel).

Chapter 8

8.1 If both strands were transcribed, each gene could encode two different polypeptides.

8.2 The −35 (RP binding site) and −10 (RP initiation site) promoter sequences and a terminator.

8.3 The promoters could have different consensus sequences for RP binding which could be differentially recognised by either (1) molecules of RP core enzyme associated with *different sigma factors* (this situation is known in *Bacillus subtilis*) or (2) *different species of RP* (for example, some phages encode their own RP, which is quite specific for the phage promoters).

8.4 If the transcripts were long-lived, it would not be possible to regulate gene activity by controlling the rate of mRNA synthesis. On the other hand, it is more economical for the tRNA and rRNA to be long-lived.

8.5 mRNA exists free in the cell and is degraded by nucleases specific for single-stranded RNA. Both tRNA and rRNA are partially double-stranded and so protected from nuclease attack. Furthermore, rRNA does not exist free, as it is normally complexed with protein, forming the ribosomes.

8.6 (1) It is charged with a formylated amino acid (*N*-formylmethionine).
(2) It is the only tRNA to respond to a codon (AUG) exposed within a 30 S ribosomal sub-unit–mRNA complex.

8.7 32.

8.8

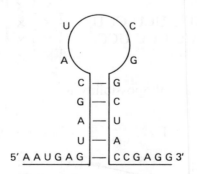

8.9 mRNA is the template against which the amino acids are assembled into polypeptides. tRNA molecules are the adaptors which recognise, on the one hand, a specific amino acid, and, on the other hand, a specific mRNA codon.

8.10 There is, for each amino acid, a specific aminoacyl tRNA synthetase, an enzyme which recognises its own amino acid and the correct species of uncharged tRNA. It joins the carboxyl group of the amino acid to the terminal nucleotide of the . . . C C A 3′ tRNA sequence in an ATP-dependent reaction.

Chapter 9

9.1 $4 \times 4 \times 4 \times 4 = 256$

9.2 The genetic code is read in groups of three from a fixed starting point. Depending on where translation commences, a nucleotide sequence can be read in three different ways — each of these is a reading frame

$$C A G * C A G * C A G * C A G *$$

or $$A G C * A G C * A G C * A G C *$$

or $$G C A * G C A * G C A * G C A *$$

9.3 Because most frameshift mutations generate an in-phase chain termination codon.

9.4 None. The strands of polyA and polyU are base-complementary and will form a duplex molecule.

9.5 Poly-methionine (reading the message as A U G * A U G * . . .) or poly-aspartic acid (reading the message as G A U * G A U * . . .). There are only two different peptides, since UGA is a chain termination codon.

9.6

	frequency among all triplets	relative frequency
UUU	$\frac{1}{2} \times \frac{1}{2} \times \frac{1}{2} = 8/64$	8
UUC UCU CUU	$\frac{1}{2} \times \frac{1}{2} \times \frac{1}{4} = 4/64$ (for each triplet)	4 (each)
CCU CUC UCC	$\frac{1}{2} \times \frac{1}{4} \times \frac{1}{4} = 2/64$ (for each triplet)	2 (each)
GGG	$\frac{1}{4} \times \frac{1}{4} \times \frac{1}{4} = 1/64$	1

9.7 The only possibility is

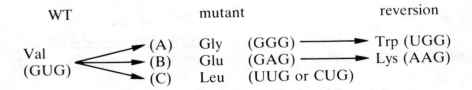

WT   mutant   reversion

Val (GUG) → (A) Gly (GGG) → Trp (UGG)

(B) Glu (GAG) → Lys (AAG)

(C) Leu (UUG or CUG)

9.8 Recombination between the DNA molecules encoding either the GAG and UUG (strains A and C) or the GGC and UUG (strains B and C) codons will generate a DNA molecule encoding the wild type GUG codon.

9.9 The only possible amino acid substitutions are leucine (mutation to UUG or CUG), valine (GUG), threonine (ACG), lysine (AAG), arginine (AGG) and isoleucine (AUU, AUC and AUA).

9.10 The altered tRNA, although charged with alanine, still has the cysteine anticodon and still responds to the cysteine codons in polyUG. Since polyUG does not normally stimulate the incorporation of alanine by alanyl-tRNA, the amino acid inserted in response to a codon **must** depend on the anticodon in the tRNA and not on the amino acid with which it is charged.

Chapter 10

10.1 In negative control the protein product of a regulator gene is a repressor of gene activity, whereas in positive control it is an activator.

10.2 Each of the structural genes has its own promoter and a common operator sequence.

10.3 (a) Mutations within a promoter resulting in increased (up) or decreased (down) levels of transcription.
(b) Relative to the promoter, downstream sequences are located in the direction of transcription. Upstream sequences are in the opposite direction.

10.4 Operator and promoter mutants are *trans*-recessive because they are control sequences, only regulating the expression of adjacent genes on the same molecule of DNA. Repressor genes encode diffusible gene products and so can influence gene expression in both *cis* and *trans*.

10.5 (a) Mutations in a structural gene (for example, $lacZ^-$).
Promoter negative mutants ($lacP^-$).
Mutations in a repressor gene resulting in a super-repressor ($lacI^s$).
(b) Mutations in an operator ($lacO^c$).
Mutations resulting in the absence of repressor ($lacI^-$).

10.6 (a) Tryptophan-requiring (there is no *trpA* gene product).
(b) Constitutive tryptophan-independent.
(c) Constitutive tryptophan-independent.

10.7 (1) The *trpOP* region is at the *trpE* end of the operon.
(2) The direction of transcription is trpEDBCA.

$\longrightarrow$

10.8 A lac^+ inducible strain cannot grow on melibiose at 42 °C, since the *lac* permease cannot be induced (neither lactose nor melibiose is present within the cells). Any constitutive mutations to $lacO^c$ or $lacI^-$ will be able to grow, since the *lac* operon is permanently expressed and melibiose can be transported into the cell by *lac* permease.

10.9 Temperate phages encode a specific repressor (the *c* gene product) which can bind to operator sites and prevent the expression of all other phage genes. In a lysogen the repressor is produced and will bind to operator sites not only on the prophage, but also on a closely related superinfecting phage — this blocks the growth of the superinfecting phage. Unrelated phages (which have operator sites sensitive to a different repressor) or phages with an operator constitutive mutation (whose operators are repressor-insensitive) can grow on lysogens.

10.10 In the absence of glucose, the levels of cAMP rise, the CAP protein binds to the CAP site within the promoter of each glucose-sensitive operon and transcription is co-ordinately initiated. In the presence of glucose, the cAMP level falls, CAP can no longer bind and the rate of transcription is co-ordinately reduced.

Chapter 11

11.1 A forward mutation is the result of a change in the nucleic acid sequence occurring more or less at random within a gene (but remember that some mutations are silent). Back mutation precisely restores the wild type nucleic acid sequence and must involve a particular type of point mutation at *one specific site* within the gene.

11.2 The amber mutations are mapped within the gene and the polypeptide produced by each mutant is examined. If the gene and its polypeptide are colinear, then the mutant mapping nearest the end of the gene at which translation is initiated will produce the shortest polypeptide and the mutant nearest the other end of the gene will produce the longest polypeptide. Hence, the peptide fragments produced by a series of mutants will form a hierarchical series and the order of the mutant sites on the genetic map will correspond to the increasing size of the polypeptides they produce.

This experiment was carried out by Sydney Brenner in 1964.

11.3 $2 \times 50\,000 \times 5 \times 10^{-5} = 5$

11.4 In intergenic suppression of missense the anticodon of a tRNA is altered, so that an acceptable amino acid is inserted in response to the missense codon. The *trpA36* mutant of *E. coli* is the result of an arginine (AGA) for glycine (GGA) substitution within the *trpA* protein. It can be suppressed by a mutation within the gene for one of the species of glycine tRNA; this mutation enables the glycine tRNA to recognise the AGA codon and to insert glycine.

This type of suppression is only weakly effective, as only a small percentage of the *trpA* proteins are active — most still have arginine at the position of the substitution.

11.5 A tautomer is one of the four bases where a hydrogen atom has changed position; the rare transient tautomeric form (frequency about 10^{-4}) can form non-standard base pairs. When a rare tautomer is incorporated into DNA, it can produce a non-standard base pair, such as A*–C (where A* is the rare tautomer of A). At the next replication normal base pairing is resumed, producing two daughter molecules, one with an A–T (wild type) and the other with a C–G (mutant) base pair.

11.6 Acridines induce frameshifts which frequently result in the premature termination of translation by generating in-phase nonsense codons. Base analogues induce base substitution type mutations, many of which generate same-sense or acceptable missense mutations.

11.7 (a) GC $\rightarrow$ AT: (b) no; (c) some (only mutations caused by AT $\rightarrow$ GC transitions); (d) yes, all.

11.8 Only proteins can be temperature-sensitive while amber mutations exert their effect during translation. Thus, these mutations are only found in structural genes encoding proteins (ii and iv).

11.9 Back mutation restores the exact nucleotide sequence present in the wild type gene. Reversion is any mutation that restores or partially restores the wild phenotype. Second site reversion is a second mutation within the mutant gene which restores the wild phenotype but not the original nucleotide sequence.

11.10 Mutation frequency is the frequency with which a particular mutation or mutant is found in a population of cells or individuals. It might, for example, be expressed as the number of mutants recovered per 10^8 bacteria plated. Mutation rate measures the probability of occurrence of a specific mutation over a given period of time. It may be expressed as the number of mutations per specific gene per generation or per cell division. In general, mutation rates are difficult to measure.

Chapter 12

12.1 The 5 S rRNA.

12.2 The 5.8 S rRNA is transcribed as part of the large molecule of pre-rRNA.

12.3 (a) Since the promoters for RPIII lie within the gene, a change in the promoter sequence will alter the nucleotide sequence of the pre-tRNA. (b) Probably none, since the tRNA genes are redundant.

12.4 Constitutively expressed genes that apparently provide the basic functions required by all cell types at all stages of development.

12.5 Incorrect splicing would disrupt the coding sequence of the gene (see Question 12.6).

12.6 The exon sequence between the two introns would be deleted.

12.7 No, it is probably an evolutionary relic. However, it is possible that new genes have been created during evolution by reassembling exons in new combinations; this hypothetical process is called **exon shuffling**.

12.8 This question is answered in Section 12.2.

12.9 (1) Initiation of transcription (no pre-mRNA).
(2) Addition of polyA tail (no pre-mRNA).
(3) Splicing (no m-RNA or altered mRNA).
(4) Initiation of translation (no polypeptide).
(5) Termination of translation (extended polypeptide).
(6) Change in coding sequence (amino acid substitution in polypeptide or a prematurely terminated polypeptide).

Chapter 13

13.1 DNA made from purified eucaryotic mRNA using reverse transcriptase. This enzyme uses RNA as a template and catalyses the first step in a multistep process producing a double-stranded DNA molecule. cDNA is a copy of mRNA and, unlike the original gene, does not contain introns; it is an uninterrupted coding sequence.

13.2 Both *Eco*R1 and *Pst*1 have six bp recognition sites and, on the average, each site will occur once in every 4^6 or 4096 bp. Thus, if the bases are randomly distributed, there will be 48.6/4.1, or approximately 12, sites for each endonuclease. The observation varies from the expectation, suggesting that the bases are not distributed at random.

13.3 The palindromic sequence 5′ CATATG 3′.

13.4 A short DNA segment about 10 bp long and containing the *Eco*R1 recognition sequence is chemically synthesised and ligated to *both* ends of the fragment. If the fragment is now treated with *Eco*R1, then *Eco*R1-specific single-stranded termini will be generated. This fragment can now be inserted into any *Eco*R1 restriction site.

13.5 (**a**) Select ampicillin-resistant transformants.
(**b**) The required clones are ampicillin-resistant and kanamycin-sensitive; any clones that are resistant to both drugs have received a religated non-recombinant plasmid.

13.6 A hybrid plasmid containing both bacterial plasmid and eucaryotic sequences, including both origins of replication. It contains markers which can be selected in either bacterial or eucaryotic cells and so can be readily transferred from one to the other (shuttled back and forth).

Index